Field Guide to the
Bats of the Amazon

2nd edition

Adrià López-Baucells, Ricardo Rocha,
Paulo Bobrowiec, Enrico Bernard,
Jorge Palmeirim and Christoph F. J. Meyer

"… foi então que o Jurupari pôs fogo e breu pra ferver, e quando ferveram, soltaram fumaça, de onde saíram morcegos, jacamins, uakuraus, murucututus, iakurutus, andorinhas e gaviões…"

"… it was then that Jurupari put fire and pitch to burn, and when it burned, bats, trumpeters, owls, swallows, hawks and other nocturnal birds emerged from the smoke…"

Legend adapted by Mônica Rodrigues da Costa,
Paula Medeiros de Oliveira and Paulo Pedro Costa

Published by Pelagic Publishing
www.pelagicpublishing.com
PO Box 725, Exeter EX1 9QU, UK

ISBN 978-1-78427-165-7 (Pbk)

Collaborating institutions

Centre for Ecology, Evolution and Environmental Changes (cE3c), Faculdade de Ciências, Universidade de Lisboa, Campo Grande C2, 1749-016, Lisboa, Portugal

Granollers Museum of Natural Sciences c/Palaudàries 102, 08402 Granollers, Catalonia

Biological Dynamics of Forest Fragments Project (BDFFP), Av. André Araujo 2936, CEP 69083-000, Manaus, Brazil

School of Environment & Life Sciences, University of Salford, Peel Building, Salford, Manchester M5 4WT, United Kingdom

Centro de Ciências Biológicas, Universidade Federal de Pernambuco, Av. Professor Moraes Rego, CEP 50670-901, Recife, Brazil

Metapopulation Research Centre, Faculty of Biosciences, University of Helsinki, Viikinkaari, FI-00014 12 Helsinki, Finland

Illustrations: Blanca Martí de Ahumada (www.blancamarti.com) & Eva Sánchez Gómez (www.evasanchez.cat)

Design: Adrià López-Baucells (www.adriabaucells.com)

Images: Oriol Massana Valeriano & Adrià López-Baucells

Cover image: *Carollia perspicillata* Copyright © Oriol Massana Valeriano & Adrià López-Baucells

Printed and bound in India by Replika Press Pvt. Ltd.

CONTENTS

PREFACE

This book is designed as a guide aimed at satisfying the needs of those conducting field work on bats in the Amazon. It is largely based on Lim et al. (2001), with modifications derived from both personal observations and three years of field experience in the Brazilian Amazon at the Biological Dynamics of Forest Fragments Project (BDFFP), as well as a thorough revision of available bat keys and scientific papers describing new species.

Our aim was to write a straightforward, easy-to-use guide that would be both practical and very visual, and would facilitate bat species identification in the field. We have tried to avoid as much as possible confusing features such as fur colour, as well as certain skull and teeth characteristics that cannot be easily assessed under field conditions.

We decided to group together many of the cryptic species that are still indistinguishable in the field and that can only reliably be identified using molecular methods such as DNA barcoding. Taxonomic nomenclature throughout this key follows Nogueira et al. (2014).

We will be delighted to receive readers' comments and suggestions!

Please send them to: adria.baucells@gmail.com

Thank you!
The Authors

Lim, B.K. & Engstrom, M.D. (2001). Species diversity of bats (Mammalia: Chiroptera) in Iwokrama Forest, Guyana, and the Guianan subregion: implications for conservation. *Biodiversity & Conservation* 10(4):613-657.

Nogueira, M.R. et al. (2014). Checklist of Brazilian bats, with comments on original records. *Check List* 10(4):808-821.

FOREWORD

This *Field Guide to the Bats of the Amazon* is the culmination of an almost unimaginable amount of challenging fieldwork. The first publication of its kind, it is beautifully illustrated, comprehensive and extraordinarily easy to use. Authors Adrià Lopez-Baucells, Ricardo Rocha, Paulo Bobrowiec, Enrico Bernard, Jorge Palmeirim and Christoph Meyer have provided an invaluable contribution to the world of bats, a must-have publication for anyone working on bats in the neotropics.

The Amazon basin encompasses more than half of our planet's remaining rainforests and is home to the world's largest, most diverse assemblage of bats. Understanding these animals is vital to the conservation of the Amazonian biome. However most Amazonian bats remain unstudied and our lack of ability to reliably identify them has been a major hindrance to research on their unique contributions and needs.

As noted, throughout this vast system, bats are essential seed dispersers, pollinators and controllers of vast numbers of herbivorous insects. Only one, the common vampire, causes significant problems for people and their livestock. Yet, far too often, all species are mistakenly killed as vampires, posing an enormous threat to the health of the whole ecosystem and associated human economies.

It is my hope that this outstanding field guide will open the door to an explosion of much needed research and education, essential to the authors' conservation goals. As a fellow photographer and conservationist I deeply appreciate the obvious attempt to show bats, even the vampires, with pleasant expressions that do not contribute to further misunderstanding and fear.

This publication is also the first to share a broad, well-organized echo-location database and key, accompanied by appropriate cautionary advice and documentation. Hopefully, it will become a model, inspiring additional field guides for the rich and also vitally important bat faunas of Africa and Asia.

Merlin D. Tuttle
Founder and Executive Director
Merlin Tuttle's Bat Conservation

ACKNOWLEDGEMENTS

This guide could not have been written without the help of the following contributors:

William Magnusson was a great help during the final stage of the production of this book. He provided us with the logistic means to publish the first edition as an online e-book, thus enabling us to widely publicize it among scientists, students and institutions. John Rose and Leon Vlieger brought Pelagic Publishing and the authors into contact.

José Luís C. Camargo, Ary Jorge C. Ferreira, M. Rosely C. Hipólito, Alaércio dos Reis, Luiz de Queiroz, Josimar Menezes, Osmaildo da Silva, and José Tenaçol provided continuous support during our fieldwork at the BDFFP.

Carles Flaquer Sánchez, Xavier Puig-Montserrat, and Antoni Arrizabalaga gave vital support to the project, and provided material and field equipment, acoustic and mist-netting training, software licenses, and valuable corrections on previous drafts.

Maria Mas helped greatly with the analysis of bat calls. Michel Barataud, Vincent Rufray and Thierry Disca are acknowledged for their help in reviewing the acoustic key.

Madalena Boto gave advice on photography and she deserves special credit for her amazing contribution to the making, montage and directing of the video trailer used to promote this guide. This video would have been impossible without the altruistic contribution from 'Of Monsters and Men' and Helena Mata from Universal Music Portugal and Syncsongs Music Publishing, who provided the instrumental piece 'Dirty Paws', used as the video soundtrack.

We would especially like to thank all the people who selflessly contributed good quality pictures to cover some of the gaps, including Merlin Tuttle, who kindly agreed to write a foreword to the book and also provided some excellent bat photographs.

Marta Acácio, Diogo Ferreira, Fabio Farneda, Gilberto Josimar, Madalena Boto, Milou Groenenberg, Júlia Treitler, Rodrigo Marciente, Solange Farias, Kevina Vulinec, Inês Silva, Joana Carvalho, Leonardo Oliveira, Ileana Mayes, and Ubirajara Capaverde were all great help in the field. They were also excellent company during these years and helped keep this project on track.

Mercè Baucells, Josep Anton López, Míriam López Baucells, Pilar Valeriano, and Ramon Massana provided financial and logistical support during the whole period.

We gratefully acknowledge institutional support from the **Centre for Ecology, Evolution and Environmental Changes (cE3c)**, the **Granollers Museum of Natural Sciences**, the **National Institute for Amazonian Research (INPA)**, the **Biological Dynamics of Forest Fragments Project (BDFFP)**, the **University of Salford**, the **Universidade Federal de Pernambuco** and the **Smithsonian Tropical Research Institure (STRI)**.

Funding was provided by a Portuguese Foundation for Science and Technology (FCT) project grant PTDC/BIA-BIC/111184/2009, SFRH/BD/80488/2011 and PD/BD/52597/2014.

PHOTOGRAPHIC CREDITS

Most of the photographs used in this field guide were taken by **Oriol Massana Valeriano** and **Adrià López-Baucells** at the Biological Dynamics of Forest Fragments Project near Manaus (Brazil) during a research project on the effects of forest fragmentation on bats undertaken in 2011–2015.

External contributions

Burton Lim, Alex Borisenko (*Lasiurus atratus* pp. 93, 149, *Thyroptera wynneae* p. 147)

Elizabeth Clare (*Eptesicus furinalis* pp. 95, 149, *Molossus sinaloae* p. 151, *Myotis albescens* pp. 97, 150, *Noctilio leporinus* p. 147, *Pteronotus davyi* p. 148, *Vampyrum spectrum* pp. 41, 51, 146).

Enrico Bernard (*Diclidurus ingens* pp. 85, 148, *Lonchorhina aurita* pp. 45, 53, 145, *Eumops bonariensis* p. 151, *Peropteryx leucoptera* pp. 83, 149, *Rhogeessa hussoni* p. 150).

Fabio Falcão (*Diaemus youngi* pp. 21, 142, *Diclidurus albus* pp. 85, 148, *Diphylla ecaudata* pp. 21, 142, *Mimon bennettii* p. 146, *Nyctinomops laticaudatus* pp. 101, 152, *Platyrrhinus lineatus* p. 143).

Fábio Z. Farneda (*Anoura geoffroyi* p. 142, *Eptesicus diminutus* p. 149, *Molossops temminckii* pp. 101, 109, *Natalus macrourus* pp. 15, 152).

Gianfranco Gómez (*Noctilio leporinus* pp. 14, 73).

Jose Gabriel Martinez Fonseca (*Choeroniscus godmani* pp. 29, 142, *Cynomops greenhalli* pp. 109, 151, *Cyttarops alecto* pp. 81, 85, 148, *Eumops perotis* p. 151, *Enchisthenes hartii* pp. 31, 37, 143, *Glossophaga commissarisi* p. 142, *Lichonycteris obscura* pp. 27 & 142, *Molossus pretiosus* pp. 105, 151).

Bruce J. Hayward (*Glossophaga longirostris* p. 142).

Lizette Siles (*Eptesicus chiriquinus* pp. 95, 149, *Nyctinomops macrotis* p. 152, *Platyrrhinus infuscus* p. 143, *Sphaeronycteris toxophyllum* pp. 31, 35, 144)

Maël Dewynter (*Lonchorhina inusitata* pp. 53, 145, *Micronycteris brosseti* p. 145, *Micronycteris minuta* p. 145, *Natalus tumidirostris* pp. 111, 152, *Phyllostomus latifolius* p. 146).

Marco Mello (*Tonatia bidens* p. 146, *Mimon bennettii* p. 51, *www.marcomello.org*).

Marisol Hidalgo-Cossio (*Peropteryx pallidoptera* pp. 87, 149)

Merlin Tuttle (*Artibeus amplus* pp. 39, 143, *Diclidurus isabella* pp. 85, 148, *Lasiurus cinereus* pp. 93, 150, *Lophostoma schulzi* p. 145, *Molossops temminckii* p. 151, *www.merlintuttle.org*).

Mónica Díaz (*Myotis simus* p. 97).

Octavio Jiménez Robles (*Eptesicus andinus* pp. 95, 149, *Eumops glaucinus* p. 151, *Molossus currentium* p. 151, *Myotis simus* p. 149, *Sturnira magna* p. 144, *Dermanura anderseni* p. 143, *Platyrrhinus infuscus* p. 143).

Ricardo Rocha (*Pteronotus personatus* p. 148).

Roberto Leonan M. Novaes (*Carollia benkeithi* p. 147, *Diclidurus scutatus* pp. 81, 85, 148, *Eumops perotis* p. 107, *Glyphonycteris sylvestris* p. 145, *Lionycteris spurrelli* p. 142, *Micronycteris schmidtorum* p. 145, *Neoplatymops mattogrossensis* p. 152, *Nyctinomops aurispinosus* p. 152, *Peropteryx macrotis* p. 149, *Peropteryx trinitatis* p. 149, *Platyrrhinus brachycephalus* p. 143, *Platyrrhinus fusciventris* p. 143, *Platyrrhinus incarum* p. 143, *Promops nasutus* p. 152, *Saccopteryx canescens* p. 149, *Scleronycteris ega* pp. 23, 27, 142, *Vampyressa pusilla* p. 144, *Vampyrodes caraccioli* pp. 41, 144).

Tiago Marques (*Centronycteris centralis* pp. 85, 148, *Dermanura glauca* p. 143)

Ubirajara Dutra (*Rhinophylla fischerae* p. 147, *Thyroptera devivoi* p. 147).

Vinícius Cardoso (*Histiotus velatus* pp. 91, 149, *Molossus coibensis* pp. 105, 151, *Thyroptera lavali* p. 147).

William Douglas de Carvalho (*Molossops neglectus* pp. 109, 151).

Vampyressa bidens

Introduction

Although elusive due to their mostly nocturnal behaviour, bats (order Chiroptera, from the Greek *cheir* 'hand' and *pteron* 'wing') are undoubtedly one of the most fascinating faunal groups in the world. Only outnumbered by rodents, they constitute the second most numerous mammalian order, but are arguably the most diverse and demonstrate just how ecologically adaptive mammals can be.

At present, over 1,300 species of bats are known to science. Nevertheless, this number is growing steadily, mostly due to the splitting of taxa based on new genetic evidence and the discovery of hitherto truly unknown species in remote corners of the planet. Bats range in size from one of the smallest of all mammals, the bumblebee bat *Craseonycteris thonglongyai* (1.5–2 g), to the large *Pteropus* flying foxes, which possess a wide array of shapes and colours; in some cases, they weigh over 1 kg and have wingspans exceeding 1.5 m. Bats have been around for some 50 million years and have taken advantage of two unique aspects of their biology – echolocation and powered flight – to conquer the night skies in nearly all of the available ecosystems across the globe, the exception being the Arctic, Antarctic and a few isolated oceanic islands.

No other mammalian order exploits such a broad diversity of food resources. Although most bat species have evolved as highly specialized hunters of aerial insects, a number have developed a taste for vertebrates (ranging from fish to amphibians, reptiles, birds and even small mammals, including other bats), plant matter (chiefly fruit, but also nectar, pollen, and occasionally leaves and seeds) and blood. Certain species are omnivorous but many bats have highly specialized diets and are involved in complex co-evolutionary interactions. A good example of this is the relationship between the South American plant *Centropogon nigricans* and its (probably) only pollinator, the recently discovered tube-lipped nectar bat *Anoura fistulata*, holder of the record for longest tongue (8.5 cm) in relation to body

Introduction

size in any mammal (its tongue measures 150% of the size of its overall body length!). Predator–prey interactions are equally intrincate and reach their evolutionary climax in the 'arms race' between aerial insectivorous bats and their prey.

Roost selection is another example of the enormous plasticity displayed by bats. Caves are probably the best-known bat roost sites; indeed many species are mostly cave-dwellers and some caves harbour millions of conspecific bats, as in the case of the Brazilian free-tailed bats *Tadarida brasiliensis* in Central America and the southern USA. Apart from caves however, bats make use of a myriad of natural and man-made structures for roosting. Some species of neotropical stenodermatine fruit-eating bats make tents by biting the central rib of palms and *Heliconia* leaves. In an interesting case of convergent evolution, *Thyroptera* bats from Central and South America and *Myzopoda* from Madagascar have both evolved suction cups or suckers on the base of their thumbs and ankles that allow them to cling to smooth surfaces and roost inside curled leaves. Some species, such as the hoary bat *Lasiurus cinereus*, are solitary tree dwellers, whilst others including many Old World fruit bats roost in large tree colonies numbering several thousands. Man-made structures such as mines, bridges and roof cavities are used by many species, while others (e.g. several neotropical Emballonuridae) simply take advantage of their camouflage to roost on lichen-covered tree bark or rocks. A few species roost in underground

cavities, while the South and Central American white-throated round-eared bat *Lophostoma silvicolum* even roosts colonially inside the nests of arboreal termites.

True powered flight and echolocation undoubtedly lie at the heart of this group's evolutionary success. Flying is much less energy-consuming than running and, given that it removes the need to touch ground, it reduces potentially deadly encounters with terrestrial predators. Echolocation probably evolved hand-in-hand with flight and, by allowing early bats to analyse the echoes of emitted sound pulses and so negotiate obstacles, served as an entrance to an ecological niche that was inaccessible to most other groups: the night sky.

Although other animal groups, including cetaceans, use sound in this way, none does so in such a complex manner. Echolocation has reached its evolutionary peak in bats and, for most species, is key to their ability to avoid physical obstacles and find food. Bats tend to have good auditory sensitivity and therefore can listen to sounds made by moving prey or, as in the case of the neotropical fringe-lipped bat *Trachops cirrhosus*, can even identify edible frogs from their calls. Good night vision and a well-developed sense of smell are also of utmost importance and enable many species to find food; this is especially true for the Old World fruit bats.

Introduction

Bats have unfortunately been the subject of disdain and persecution by many, and are frequently portrayed as blood-sucking demons and associated with dark practices. On the other hand, some cultures such as the Middle-to-Late Qing Dynasty (1644–1911) in China have regarded bats as symbols of good fortune, a much more faithful reflection of their importance to the planet's ecological health and to our own well-being. Bats are key providers of many ecosystem services such as seed dispersal, pollination and pest suppression. Their disappearance can lead to enormous economic losses (e.g. the economic value of bats to North American agriculture alone has been estimated at around $23 billion per year) and probable wide-scale ecosystem collapse.

Over the last 500 years the planet has faced a human-generated wave of extinctions that is comparable to the Earth's five previous mass extinctions. Despite their uniqueness, bats face the same threats as many other species on the planet and are consequently being severely affected by the ongoing 'sixth mass extinction'. Currently, approximately one-quarter of all bat species are globally threatened. Increasing rates of habitat loss and fragmentation, overexploitation, misguided persecution, climate change, and epidemic diseases (such as white-nose syndrome, a fungal infection that has killed millions of bats throughout North America in recent years) mean that many more species are likely to become extinct in the near future.

Fortunately not all is gloom. As we come to better understand bats, their importance for ecosystem well-being and functioning, and ultimately, how they benefit humankind, attitudes towards them are slowly starting to change. Across the globe multiple grass-roots conservation projects are braving their way to try to reverse ongoing population declines and the image of bats in books, movies and the general media is starting to reflect some elements of truth. Conservation of the planet's unique biological richness will ultimately depend on how much we treasure the natural world. We hope that by revealing some of the tremendous richness of

Lonchophylla thomasi

the Amazonian bat fauna this book will aid in a better understanding of their natural history, our impacts on them and, consequently, how we can combine our efforts to better contribute to their conservation, because as the Senegalese conservationist Baba Dioum once said:

> *"In the end we will conserve only what we love.*
> *We will love only what we understand.*
> *We will understand only what we are taught."*

Bats in the Amazon

The increase in species richness with increasing proximity to the Equator is a major biogeographic pattern to which bats are no exception. Bat diversity peaks in tropical regions, and the neotropics of South and Central America constitute the epicentre of this diversity, harbouring more than 200 currently recognized species.

The Amazon basin holds over half of the world's remaining rainforests and represents the largest and most biodiverse expanse of tropical rainforest on the planet. Roughly one in ten known bat species occurs in the Amazon basin and in some Central Amazonian localities more than 100 species live in sympatry.

Bats are divided into 17 families (or 18, depending on the acceptance of Miniopteridae as a separate family), of which nine (Phyllostomidae, Thyropteridae, Furipteridae, Noctilionidae, Mormoopidae, Emballonuridae, Vespertilionidae, Molossidae, and Natalidae) are present in the Amazon. The distribution of the species across the Amazonian bat families is rather uneven: the bulk of species belongs to the family of New World leaf-nosed bats (Phyllostomidae), the ecologically most diverse family within the order (nearly 200 species throughout Central and South America). On the other hand, the Furipteridae are represented in the Amazon by just one of the two members of its family, the thumbless bat *Furipterus horrens*.

Bats are key elements in the Amazon's intricate ecological networks and, through countless links to other animal and plant groups, help support and sustain the biome in all its complexity and magnificence. Many Amazonian bats such as the Phyllostomidae subfamilies Stenodermatinae and Carolliinae feed almost exclusively on fruit and act as 'forest gardeners' by dispersing seeds far and wide. They often introduce seeds into previously disturbed habitats and consequently help the forest reclaim some of its lost domains. Some other species such as the Glossophaginae hover like hummingbirds in front of flowers and with their long muzzles and tongues probe flowers to extract their nectar, effectively acting as

Lasiurus egregius

Bats in the Amazon

pollinators, thereby helping to maintain the genetic diversity of flowering plants. However, most Amazonian bats are either obligate or facultative insect-eaters and glean insects and other arthropods directly from the vegetation in the forest understory, or capture prey in open spaces above or below the forest canopy. By doing so, they greatly reduce arthropod-related herbivory and redistribute nutrients via their guano, thereby helping to maintain terrestrial and aquatic ecosystems throughout the Amazon. Four species of Phyllostomidae, namely the greater spear-nosed bat *Phyllostomus hastatus*, the fringe-lipped bat *Trachops cirrhosus*, the big-eared woolly bat *Chrotopterus auritus*, and the spectral bat *Vampyrum spectrum*, are confirmed carnivores, while the two *Noctilio* species are both fish-eaters. On the other hand, bats regularly form part of the diet of several faunal groups including spiders, giant centipedes, frogs, marsupials, other bats, birds, and snakes.

In recent years several new species have been described and new records have extended the geographic range of some species by hundreds of kilometers. However, knowledge of Amazonian bats is still limited and extremely biased towards certain relatively well-studied localities such as the Biological Dynamics of Forest Fragments Project (BDFFP) and Alter do Chão, in the heart of the Brazilian Amazon. As bat researchers venture into the last unknown Amazonian frontiers we are learning more about the fascinating diversity of the bats of this region, knowledge that is vital for both bat conservation and the conservation of the Amazon biome as a whole.

Dermanura gnoma

How to use this guide

Bat morphology and terminology used in this guide

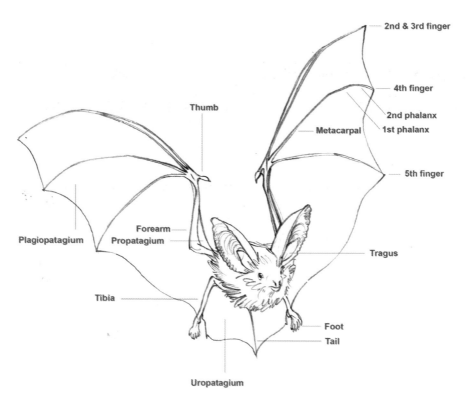

2nd & 3rd finger

4th finger

Thumb

2nd phalanx

Metacarpal — 1st phalanx

5th finger

Forearm

Plagiopatagium — Propatagium

Tragus

Tibia

Foot

Tail

Uropatagium

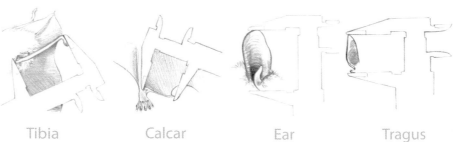

Tibia   Calcar   Ear   Tragus

How does it work?

(A) This is not a dichotomous key. Each choice may lead to a number of hierarchical options!

(B) Species that are virtually indistinguishable in the field have been grouped together. Consider collecting wing-punches for genetic studies.

(C) All measurements are given in mm.

(D) Forearm length (FA) is given after each species name. However, FA length may vary geographically and thus may not always be a reliable characteristic!

This symbol indicates that the use of a good hand-lens or camera is required.

Phyllostomidae

Thyropteridae

Furipteridae

Noctilionidae

Mormoopidae

Emballonuridae

Vespertilionidae

Molossidae

Natalidae

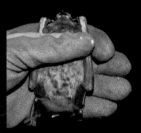

Holding a bat

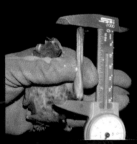

Forearm (FA)

Noseleaf

Thumb (with nail)

Key to Amazonian bat families

1a. Noseleaf or flaps of skin on face.

Phyllostomidae (p. 24)

1b. Wrists and ankles with suction cup.

Thyropteridae (p. 72)

1c. Rudimentary thumb with reduced claw almost entirely embedded in propatagium.

Furipteridae (p. 76)

1d. Tail emerges from dorsal surface of the uropatagium.

    2a. Upper lip drooping, split frontally; feet/claws very large.

Noctilionidae (p. 78)

    2b. Chin with bumps or folds of skin; upper lip not split, feet/claws not particularly enlarged.

Mormoopidae (p. 82)

    2c. Enlarged muzzle; glandular sac present in tail or forearm (sometimes vestigial in females).

Emballonuridae (p. 86)

1e. Tail enclosed and extending to the edge of pointed uropatagium.

Vespertilionidae (p. 96)

1f. Tail extending well beyond the edge of the uropatagium.

    2a. Short legs and slim wings; fur quite short and oily.

Molossidae (p. 106)

    2b. Ears large, funnel-shaped; depressed face; fur ranges from yellowish to orangish; tail equal to or longer than body length.

Natalidae (p. 118)

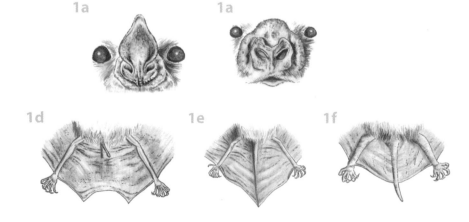

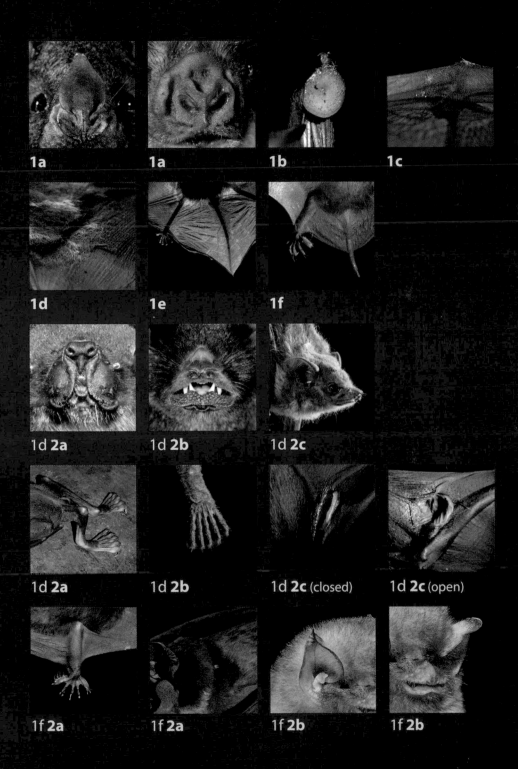

1a   1a   1b   1c

1d   1e   1f

1d 2a   1d 2b   1d 2c

1d 2a   1d 2b   1d 2c (closed)   1d 2c (open)

1f 2a   1f 2a   1f 2b   1f 2b

Phyllostomidae

The New World leaf-nosed bats constitute one of the most extraordinary examples of adaptive radiation in the natural world. The nearly 200 recognized species have most probably evolved from an insectivorous ancestor; nevertheless, although insectivory is still the predominant dietary strategy amongst Phyllostomidae, numerous species have evolved to exploit other food sources such as fruit, nectar, pollen, small vertebrates, and, in the case of the three vampire bat species, even blood.

Phyllostomids range in size from the small white-shouldered bat *Ametrida centurio* (FA averages 26 mm) to the carnivorous false vampire bat *Vampyrum spectrum* (FA averages 106 mm), the largest bat native to the Neotropics. Their morphological features are quite variable, reflecting this family's diverse diet and foraging behaviours; even so, most species have an often large, blade-shaped noseleaf, from which both the scientific and common names of this family derive. This noseleaf is thought to act as an acoustic pointer and magnifier that concentrates echolocation calls into a narrow beam.

Lophostoma silvicolum

Lophostoma silvicolum

Phyllostomidae

1a. Noseleaf greatly reduced; incisors blade-like; thumbs greatly enlarged.

Desmodontinae (p. 20)

1b. Elongated muzzle; tongue remarkably long, sometimes protruding from mouth.

Glossophaginae (p. 22)

1c. Either whitish stripes on face or head (crown) or whitish patches on shoulders or uropatagium absent.

Stenodermatinae (p. 30)

1d. Chin with warts in V/Y shape, similar in size, with no large central wart; ears often very large.

Phyllostominae (p. 44)

1e. Chin with large central and rounded wart surrounded by smaller protuberances or two enlarged bumps.

Carolliinae (p. 60)

1a

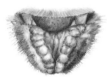

1d

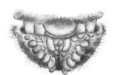

1d

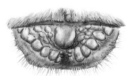

1e

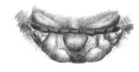

1e

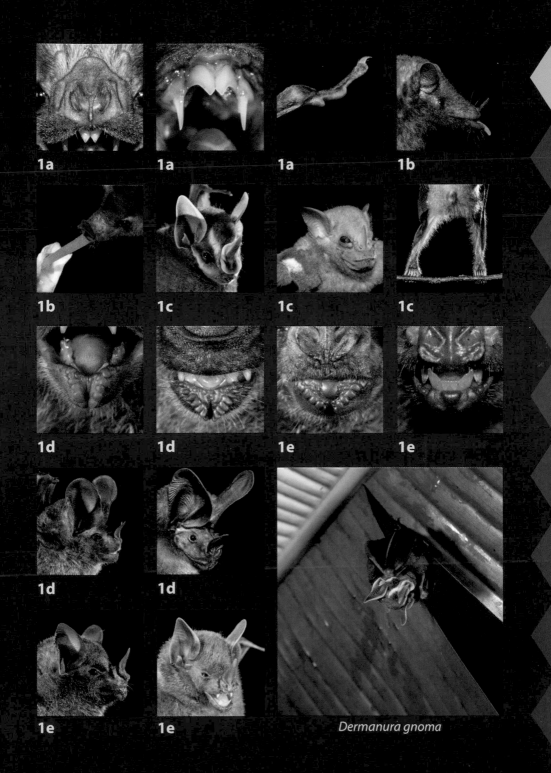

1a

1a

1a

1b

1b

1c

1c

1c

1d

1d

1e

1e

1d

1d

1e

1e

Dermanura gnoma

Phyllostomidae / *Desmodontinae*

1a.   Thumb < 13 mm; no pad under thumb.

Diphylla (p. 20)

1b.   Thumb > 13 mm; one or two pads under thumb.

2a.   Calcar absent; one long pad under thumb, whitish tips
on wings.

Diaemus (p. 20)

2b.   Tiny calcar present; two rounded pads under thumb;
darker tips on wings.

Desmodus (p. 20)

Diphylla (hairy-legged vampire bat)

1a. Only one species in the genus.

Diphylla ecaudata (50–56 mm)

Diaemus (white-winged vampire bat)

1a. Only one species in the genus.

Diaemus youngi (50–56 mm)

Desmodus (common vampire bat)

1a. Only one species in the genus.

Desmodus rotundus (52–63 mm)

1a   1b 2a   1b 2b

 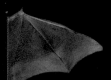 

1b   **1b 2b**   **1b 2b**

  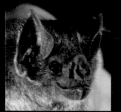

Diphylla   *Diaemus*   *Desmodus*
ecaudata   *youngi*   *rotundus*

Desmodus rotundus

Phyllostomidae / *Glossophaginae*

1a. Lower incisors absent.

2a. Uropatagium hairy and small, does not enclose the knees.

Anoura (p. 26)

2b. Uropatagium naked and encloses the knees.

3a. Dorsal fur tricoloured: dark brown-pale-dark brown.

Lichonycteris (p. 26)

3b. Fur bicoloured.

4a. 1st phalanx of thumb shorter than 2nd.

Scleronycteris (p. 26)

4b. 1st and 2nd phalanxes of thumb the same length.

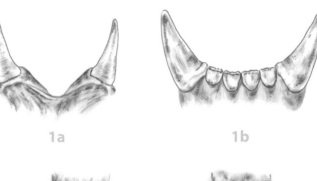

1a   1b

 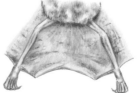

1a 2a   1a 2b

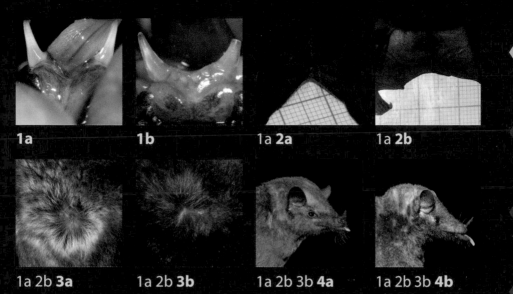

1a   **1b**   1a **2a**   1a **2b**

1a 2b **3a**   1a 2b **3b**   1a 2b 3b **4a**   1a 2b 3b **4b**

Lonchophylla thomasi

Phyllostomidae / Glossophaginae

1b. Lower incisors present (although sometimes surrounded by the gum).

2a. Upper incisors roughly similar in size, forming an arch.

Glossophaga (p. 28)

2b. Upper central incisors much larger than lateral ones.

3a. Dorsal fur strongly bicoloured; wing attached to ankle.

Lonchophylla (p. 28)

3b. Dorsal fur unicoloured; wing attached near base of toe.

Lionycteris (p. 28)

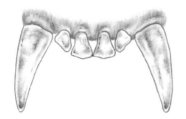

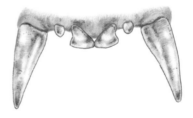

1b **2a**   1b **2b**

1b 2b **3a**   1b 2b **3b**

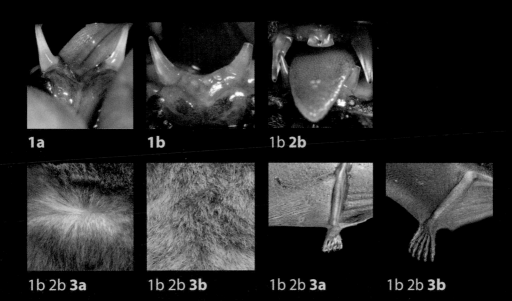

1a   1b   1b 2b

1b 2b **3a**   1b 2b **3b**   1b 2b **3a**   1b 2b **3b**

Lonchophylla thomasi

Phyllostomidae / Glossophaginae

Choeroniscus (p. 28)

Anoura (hairy-legged long-tongued bats)

1a. Tail small but present.

Anoura caudifer (34–39 mm)

1b. Tail absent; dorsal fur bicoloured; venter uniform brown.

Anoura geoffroyi (39–47 mm)

Lichonycteris (dark long-tongued bat)

1a. Only one species in the Amazon.

Lichonycteris degener (30–36 mm)

Scleronycteris (Ega's long-tongued bat)

1a *Anoura*   **1b** *Anoura*   *Lichonycteris obscura*   *Scleronycteris ega*

Lonchophylla thomasi

Phyllostomidae / *Glossophaginae*

1a.   Only one species in the genus.

Scleronycteris ega (33–35 mm)

Choeroniscus (long-nosed long-tongued bats)

1a.   Only one species complex in the Amazon.

Choeroniscus godmani (31–38 mm) / *minor* (26–39 mm)*

Glossophaga (long-tongued bats)

1a.   Lower incisors unspaced, large and weakly cusped.

Glossophaga longirostris (35–42 mm)*

1b.   Lower incisors unspaced, peg-like.

Glossophaga soricina (31–40 mm)*

1c.   Lower incisors small and medially separated by small gap.

Glossophaga commissarisi (31–38 mm)*

Lonchophylla (Thomas' nectar bat)

1a.   Only one species in the Amazon.

Lonchophylla thomasi (29–35 mm)

Lionycteris (chestnut long-tongued bat)

1a.   Only one species in the genus.

Lionycteris spurrelli (32–38 mm)

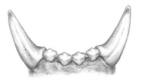
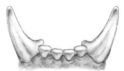
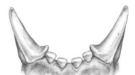

1a   1b   1c

Glossophaga

---

* We recommend classification as *Choeroniscus godmani* / *minor* and *Glossophaga* sp. until more external morphological data are available for a reliable identification in the field.

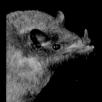

*Choeroniscus
godmani*

Choeroniscus minor

*Lonchophylla
thomasi*

Glossophaga soricina

Phyllostomidae / *Stenodermatinae*

1a. Uropatagium absent; shoulders orangish/yellowish.

                                                          Sturnira (p. 34)

1b. Uropatagium present; shoulders with white patch.

    2a. Noseleaf clearly distinct.
        ♂ *has two glands on breast* / ♀ *greatly enlarged clitoris*

                                            Ametrida (p. 34)

    2b. Noseleaf not distinct; protuberance emerging from the face.

                                Sphaeronycteris (p. 34)

1c. Uropatagium present; shoulders without coloured patch.

    2a. Inner upper incisors bifid.

        3a. Dorsal stripe present.

                                    Uroderma (p. 34)

        3b. Dorsal stripe absent.

            4a. FA < 43 mm.

                5a. Base of noseleaf joined to lip.

                              Enchisthenes (p. 36)

                5b. Base of noseleaf separate from lip.

                              Dermanura (p. 36)

            4b. FA > 43 mm.

                              Artibeus (p. 38)

    2b. Inner upper incisors not bifid.

1c **2a**   1c **2b**   1c 2a 3b   1c 2a 3b

                                                4a **5a**   4a **5b**

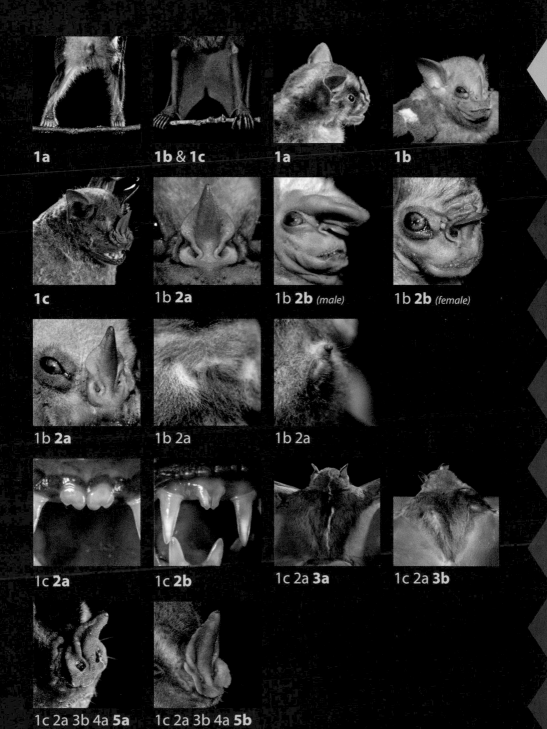

1a 1b & 1c 1a 1b

1c 1b 2a 1b 2b *(male)* 1b 2b *(female)*

1b 2a 1b 2a 1b 2a

1c 2a 1c 2b 1c 2a 3a 1c 2a 3b

1c 2a 3b 4a 5a 1c 2a 3b 4a 5b

Phyllostomidae / *Stenodermatinae*

3a. Uropatagium only furry at edges.

Vampyrodes & Platyrrhinus (p. 40)

3b. Uropatagium furry dorsally; always four lower incisors.

Chiroderma (p. 42)

3c. Uropatagium not furred; two or four lower incisors.

4a. Facial stripes present; fur dark brown.

Vampyressa & Vampyriscus (p. 42)

4b. Facial stripes very indistinct; fur pale, almost whitish.

Mesophylla (p. 42)

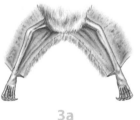

3a

3b

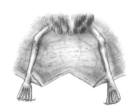

3b

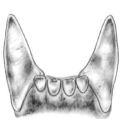

3b

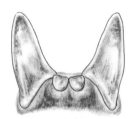

3c

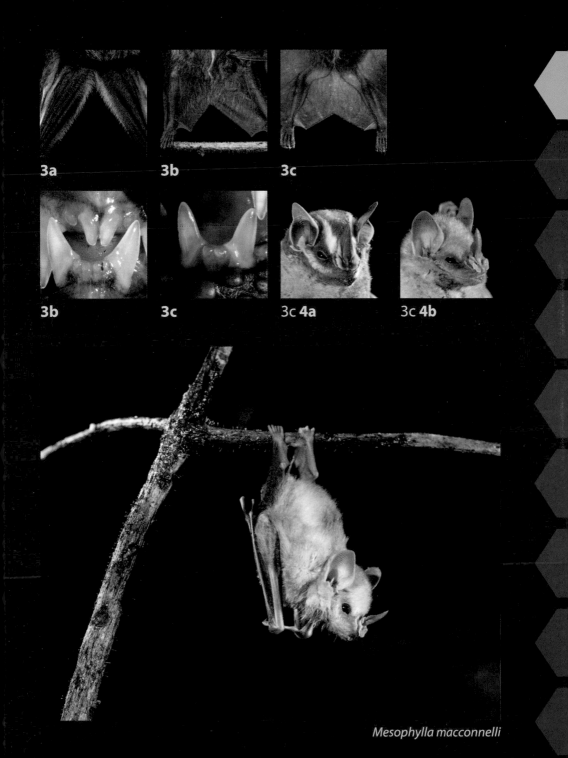

3a   **3b**   **3c**

3b   **3c**   3c **4a**   3c **4b**

Mesophylla macconnelli

Phyllostomidae / Stenodermatinae

Sturnira (yellow-shouldered bats)

1a. FA < 45 mm; inner upper incisors pointed; fur bicoloured.

Sturnira lilium (36–45 mm)

1b. FA 44-48 mm; inner upper incisors flattened; fur tricoloured.

Sturnira tildae (44–48 mm)

1c. FA > 55 mm.

Sturnira magna (55–60 mm)

Ametrida (little white-shouldered bat)

1a. Only one species in the genus.

Ametrida centurio (24–33 mm)

Sphaeronycteris (visored bat)

1a. Only one species in the genus.

Sphaeronycteris toxophyllum (37–42 mm)

Uroderma (tent-making bats)

1a. Facial stripes distinct; ears with white edges.

Uroderma bilobatum (39–45 mm)

1b. Facial stripes indistinct; ears with brownish edges.

Uroderma magnirostrum (39–45 mm)

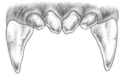
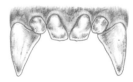

1a   1b

Sturnira

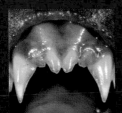 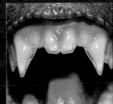 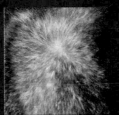 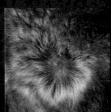

1a *Sturnira*   **1b** *Sturnira*   **1a** *Sturnira*   **1b** *Sturnira*

Ametrida centurio   *Sphaeronycteris toxophyllum* (♂)   *Sphaeronycteris toxophyllum* (♀)

  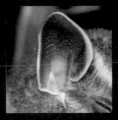 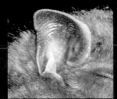

1a *Uroderma*   **1b** *Uroderma*   **1a** *Uroderma*   **1b** *Uroderma*

Sturnira tildae

Phyllostomidae / Stenodermatinae

Enchisthenes (velvety fruit-eating bat)

1a. Only one species in the genus.

Enchisthenes hartii (36–42 mm)

Dermanura (fruit-eating bats)

1a. Uropatagium furry.

Dermanura anderseni (38–40 mm)

1b. Uropatagium bare.
    2a. Facial stripes indistinct; V-shaped uropatagium.

Dermanura glauca (37–42 mm)

    2b. Facial stripes distinct; U-shaped uropatagium; ears and
        base of noseleaf with white-to-yellow edges.

Dermanura gnoma / cinerea* (34–42 mm)

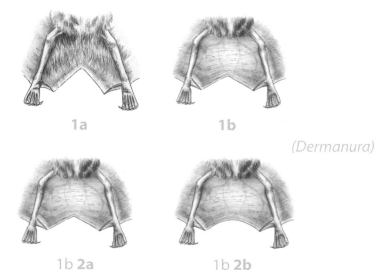

1a   1b

(Dermanura)

1b **2a**   1b **2b**

---

* We recommend classification as *D. gnomus / cinerea* until more external morphological data are available for reliable identification in the field.

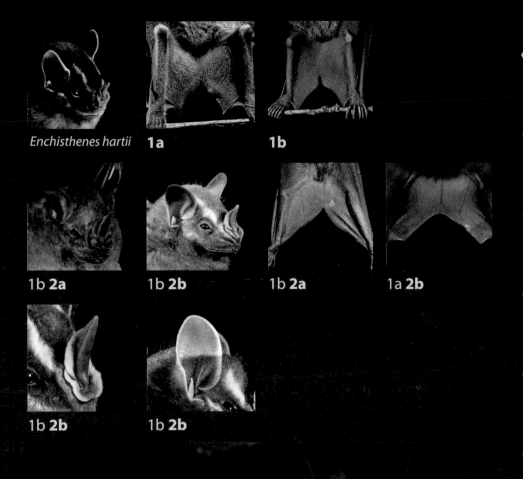

Enchisthenes hartii **1a**   **1b**

1b 2a   **1b 2b**   **1b 2a**   **1a 2b**

1b 2b   **1b 2b**

Dermanura gnoma

Phyllostomidae / Stenodermatinae

Artibeus (fruit-eating bats)

1a.   FA < 55 mm.

Artibeus concolor (43–52 mm)

1b.   FA > 55 mm.

    2a.   Facial stripes indistinct.

        3a.   Presence of a few hairs longer than fur.

Artibeus planirostris (56–73 mm)

        3b.   Absence of hairs longer than fur.

Artibeus obscurus (55–65 mm)

    2b.   Facial stripes evident.

        3a.   Uropatagium dorsally furry.

Artibeus lituratus (65–78 mm)

        3b.   Uropatagium not furry.

            4a.   Bottom of noseleaf separate from lip.

Artibeus planirostris (56–73 mm)

            4b.   Bottom of noseleaf blends with lip.

Artibeus amplus (65–74 mm)

1b 2b **3a**   1b 2b **3b**

Notice the bottom part of the noseleaf blending with the lip.

1b 2b 3b **4a**   1b 2b 3b **4b**

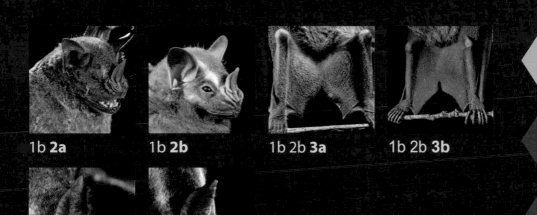

1b **2a**   1b **2b**   1b 2b **3a**   1b 2b **3b**

1b 2b 3b **4a**   1b 2b 3b **4b**

Artibeus obscurus

Phyllostomidae / *Stenodermatinae*

Platyrrhinus & Vampyrodes
(white-lined fruit bats & great stripe-faced bat)

1a. FA > 54 mm.

Platyrrhinus infuscus (54–62 mm)

1b. FA 43–55 mm.

Platyrrhinus aurarius (51–54 mm) / *Vampyrodes caraccioli* (46–57 mm)
Platyrrhinus lineatus (43–52 mm)

1c. FA < 42 mm.

    2a. V-shaped uropatagium.

Platyrrhinus fusciventris (35–40 mm)

    2b. U-shaped uropatagium.

Platyrrhinus incarum (33–42 mm) /
Platyrrhinus brachycephalus (33–42 mm)

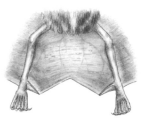 

1c **2a**   1c **2b**

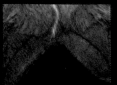 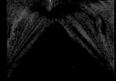 

1c **2a**   1c **2b**   *Vampyrodes*
                                                        caraccioli

Fieldwork (Central Amazon)

Phyllostomidae / *Stenodermatinae*

Chiroderma (big-eyed bats)

1a. FA > 44 mm; facial stripes faint; dorsal stripe absent.

Chiroderma villosum (44–50 mm)

1b. FA < 43 mm; facial stripes distinct; dorsal stripe present.

Chiroderma trinitatum (38–43 mm)

Vampyressa & *Vampyriscus* (yellow-eared bats)

1a. Two lower incisors.

Vampyriscus bidens (34–38 mm)

1b. Four lower incisors.

2a. Dorsal line faint.

Vampyriscus brocki (29-35 mm)

2b. Dorsal line absent; FA usually < 34 mm.

Vampyressa pusilla / thyone* (30–36 mm)

Mesophylla (Macconnell's bat)

1a. Only one species in the genus.

Mesophylla macconnelli (29–34 mm)

(Vampyressa & Vampyriscus)

1a   1b

---

* We recommend classification as *V. pusilla / thyone* until more external morphological data are available for reliable identification in the field.

  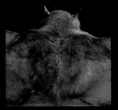 

1a *Chiroderma* **1b** *Chiroderma* **1a** *Chiroderma* **1b** *Chiroderma*

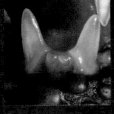 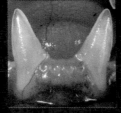  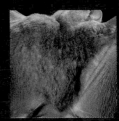

1a *Vampyressa* **1b** *Vampyressa* 1b **2a** 1b **2b**

Mesophylla
macconnelli

Mesophylla macconnelli

Phyllostomidae / Phyllostominae

1a. Well-developed protuberances on lips and chin.

Trachops (p. 50)

1b. Cup-shaped noseleaf; FA > 75 mm.

    2a. FA > 88 mm; tail absent.

Vampyrum (p. 50)

    2b. FA < 87 mm; tail present.

Chrotopterus (p. 50)

1c. Two lower incisors.

    2a. Noseleaf long and blade-shaped; furry ears.

Mimon (p. 50)

    2b. Noseleaf not as above; ears bare.

Lophostoma / *Tonatia* (p. 52)

1d. Four lower incisors.

    2a. Tail extending to the edge of the uropatagium.

        3a. FA > 40 mm; noseleaf length > 3 times its width; uropatagium pointed with no rows of dots.

Lonchorhina (p. 52)

        3b. FA < 40 mm; noseleaf length < 3 times its width; uropatagium squarish with rows of dots.

Macrophyllum (p. 54)

    2b. Tail not extending to the edge of the uropatagium.

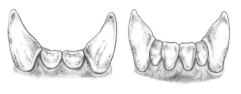

1c   1d

1d **2a**   1d **2b**

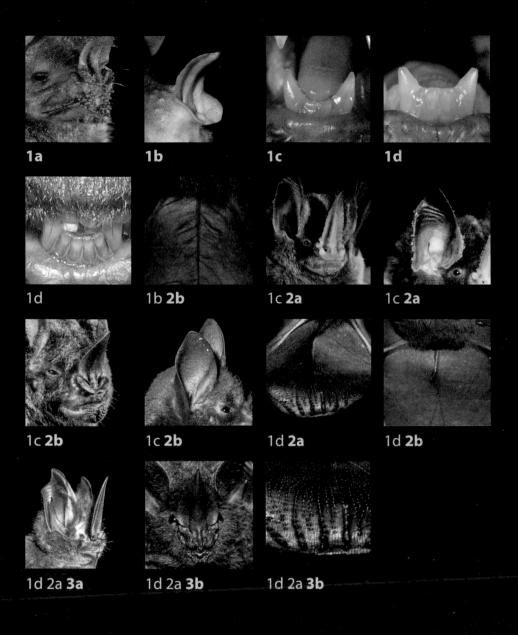

1a   1b   1c   1d

1d   1b 2b   1c 2a   1c 2a

1c 2b   1c 2b   1d 2a   1d 2b

1d 2a 3a   1d 2a 3b   1d 2a 3b

Phyllostomidae / *Phyllostominae*

3a.   FA > 58 mm; chin with flat round bumps.

  4a.   Wing tips dark; face furry.

Phyllostomus (p. 54)

  4b.   Wing tips whitish; face bare.

Phylloderma (p. 54)

3b   FA < 58 mm; chin with smooth elongated pads in V-shape.

  4a.   Length of inner upper incisors equal to canine length.

Glyphonycteris (p. 54)

  4b.   Upper incisors clearly shorter than canines.

    5a.   FA < 35 mm.

Neonycteris (p. 56)

    5b.   FA > 35 mm.

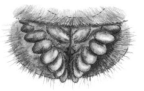

3a

3b

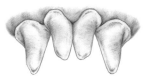

3b **4a**

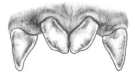

3b **4b**

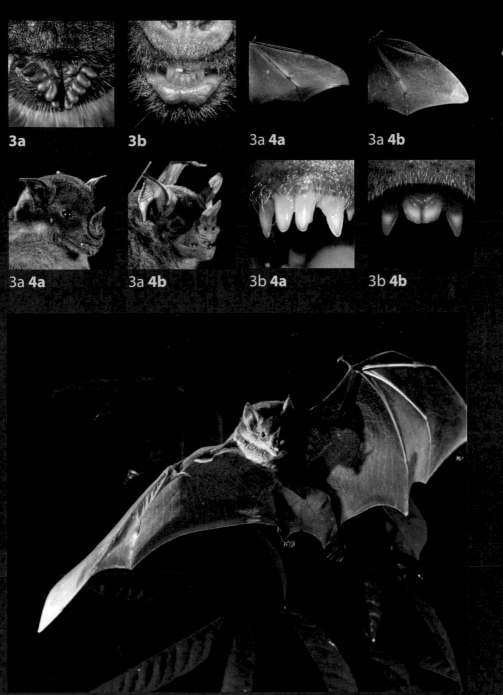

3a   **3b**   3a **4a**   3a **4b**

3a **4a**   3a **4b**   3b **4a**   3b **4b**

Phylloderma stenops

Phyllostomidae / *Phyllostominae*

Trachops (frog-eating bat)

1a. Only one species in the genus.

Trachops cirrhosus (57–66 mm)

Vampyrum (spectral bat)

1a. Only one species in the genus.

Vampyrum spectrum (88–114 mm)

Chrotopterus (false vampire bat)

1a. Only one species in the genus.

Chrotopterus auritus (77–87 mm)

Mimon (Gray's spear-nosed bats)

1a. Dorsal stripe absent; noseleaf smooth and bare; wing attached to ankle.

Mimon bennettii (51–59 mm)

1b. Dorsal stripe present; noseleaf serrated and hairy; wing attached to foot.

Mimon crenulatum (46–55 mm)

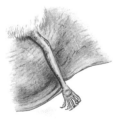 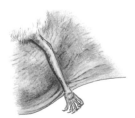

(Mimon)

1a   1b

Trachops cirrhosus *Vampyrum spectrum*

Chrotopterus auritus

1a *M. bennettii*   **1b** *M. crenulatum*

Trachops cirrhosus

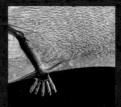

1a *M. bennettii*   **1b** *M. crenulatum*   **1a** *M. bennettii*   **1b** *M. crenulatum*

Phyllostomidae / *Phyllostominae*

Lophostoma & *Tonatia* (round-eared bats)

1a. Venter pure white.

Lophostoma carrikeri (43–50 mm)

1b. Venter pale brown to brown.

2a. FA < 49 mm.

3a. Small warts on forearm.

Lophostoma schulzi (42–45 mm)

3b. No warts on forearm.

Lophostoma brasiliense (32–36 mm)

2b. FA > 50 mm.

3a. Forearm furry; ears separate.

4a. Faint stripe between ears.

Tonatia saurophila (51–59 mm)

4a. No stripe between ears.

Tonatia bidens (48–60 mm)

3b. Forearm bare; ears connected by band.

Lophostoma silvicolum (49–60 mm)

Lonchorhina (sword-nosed bats)

1a. FA 52–57 mm; long muzzle.

Lonchorhina inusitata (52–57 mm)

1b. FA 47–54 mm; short muzzle.

Lonchorhina aurita (47–54 mm)

1a *Lop. & Ton.*   **1b** *Lop. & Ton.*   1b 2a **3a**   1b 2a **3b**

1b 2b **3a**   1b 2b **3b**   1b 2b **3a**   1b 2b **3b**

1b 2b 3a **4a**   1b 2b 3a **4b**   **1a** *Lonchorhina*   **1b** *Lonchorhina*

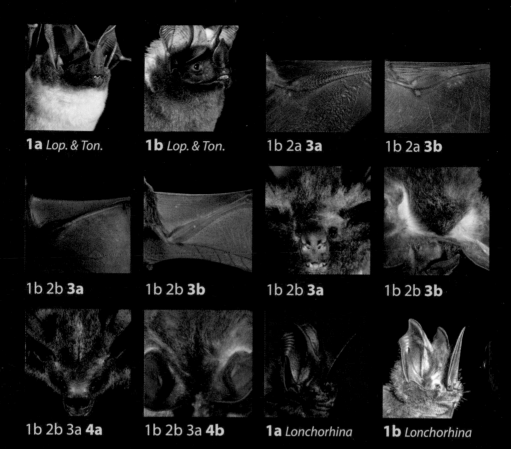

Lophostoma carrikeri

Phyllostomidae / Phyllostominae

Macrophyllum (long-legged bat)

1a. Only one species in the genus.

Macrophyllum macrophyllum (32–40 mm)

Phyllostomus (spear-nosed bats)

1a. FA > 75 mm.

Phyllostomus hastatus (77–93 mm)

1b. FA < 75 mm.

    2a. Calcar < Hindfoot.

Phyllostomus discolor (55–69 mm)

    2b. Calcar > Hindfoot.

        3a. FA 61–69 mm; tibia > 24 mm; venter dark
            with no frosting. *Phyllostomus elongatus* (61–71 mm)

        3b. FA 56–61 mm; tibia < 24 mm; venter dark
            with frosting. *Phyllostomus latifolius* (56–61 mm)

Phylloderma (pale-faced bat)

1a. Only one species in the genus.

Phylloderma stenops (65–83 mm)

Glyphonycteris (grey-bearded bats)

1a. FA > 50 mm; two upper incisors.

Glyphonycteris daviesi (52–59 mm)

1b. FA < 45 mm; four upper incisors.

Glyphonycteris sylvestris (37–44 mm)

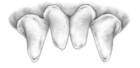

1a

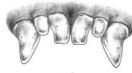

1b

(Glyphonycteris)

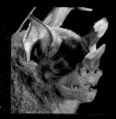

Macrophyllum macrophyllum

Phylloderma stenops

1b **2a**

1b **2b**

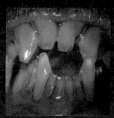

1b 2b **3a**

1b 2b **3b**

1a *Glyphonycteris*

Phyllostomus discolor

Phyllostomidae / Phyllostominae

Neonycteris (least big-eared bat)

1a. Only one species in the genus.

Neonycteris pusilla (33–35 mm)

Micronycteris (big-eared bats)

1a. Dark venter.

    2a. FA > 40 mm; lower incisors narrow.

Micronycteris hirsuta (40–46 mm)

    2b. FA < 37 mm.

        3a. Ears < 22 mm.

Micronycteris megalotis (31–36 mm)

        3b. Ears > 22 mm.

Micronycteris microtis (32–37 mm)

1b. Venter white or pale.

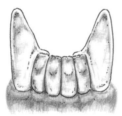

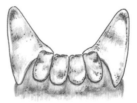

1a **2a**   1a **2b**

  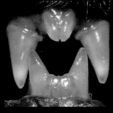 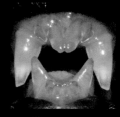

1a   **1b**   1a **2a**   1a **2b**

Micronycteris microtis

Phyllostomidae / *Phyllostominae*

2a. Calcar > Hindfoot.

    3a. FA 33–38 mm; tibia > 14.5 mm.

        Micronycteris schmidtorum (33–38 mm)

    3b. FA 31–34 mm; tibia < 14.5 mm.

        Micronycteris brosseti (31–34 mm)

2b. Calcar ≤ Hindfoot.

    3a. Digit IV: 1st > 2nd phalanx.

        Micronycteris homezorum (34–37 mm)

    3b. Digit IV: 1st = 2nd phalanx.

        Micronycteris minuta / sanborni (31–37 mm)*

Trinycteris (Niceforo's big-eared bat)

1a. Only one species in the genus.

        Trinycteris nicefori (35–41 mm)

Lampronycteris (yellow-throated big-eared bat)

1a. Only one species in the genus.

        Lampronycteris brachyotis (38–44 mm)

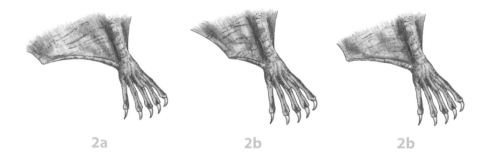

    2a   2b   2b

---

* We recommend classification as *M. minuta/sanborini* until more external morphological data are available for reliable identification in the field.

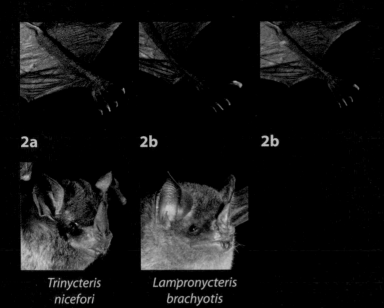

2a   2b   2b

Trinycteris
nicefori

Lampronycteris
brachyotis

Ephemeral lake in the Amazon

Phyllostomidae / Carolliinae

1a. Tail short but present; dorsal fur bi- or tricoloured.

Carollia (p. 60)

1b. Tail absent; dorsal fur unicoloured.

Rhinophylla (p. 62)

Carollia (short-tailed fruit bats)

1a. Faint banding pattern on dorsal fur; tibia 14–17 mm.

Carollia castanea / benkeithi* (34–39 mm)

1b. Clear banding pattern on dorsal fur.

    2a. Tibia 16–17 mm.

Carollia brevicauda** (27–42 mm)

    2b. Tibia 17–21 mm.

Carollia perspicillata** (38–44 mm)

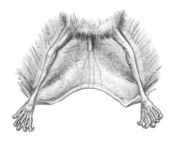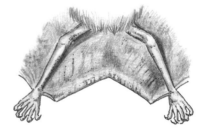

1a   1b

(Carolliinae)

---

* We recommend classification as *C. castanea / benkeithi* until more external morphological data are available for reliable identification in the field.

** We recommend classification as *C. brevicauda / perspicillata* in doubtful cases until more external morphological data are available for reliable identification in the field.

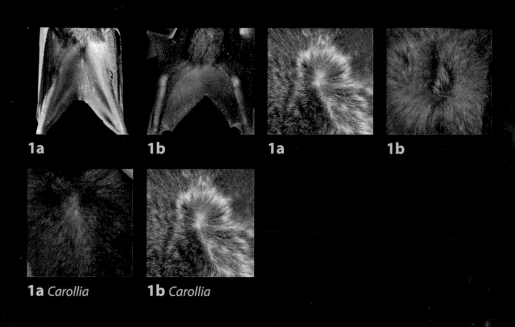

1a   **1b**   **1a**   **1b**

1a *Carollia*   **1b** *Carollia*

Carollia perspicillata

Phyllostomidae / Carolliinae

Rhinophylla (little fruit bats)

1a. Uropatagium with bare edge; legs very furry; no gap between upper incisor and canine.

Rhinophylla pumilio (33–36 mm)

1b. Uropatagium with furry edge; legs bare; gap between upper incisor and canine.

Rhinophylla fischerae (29–34 mm)

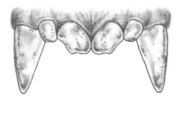

1a

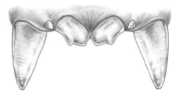

1b

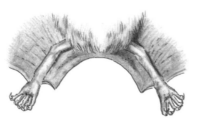

1a

1b

1a

1a

Rhinophylla pumilio

Thyropteridae *(disc-winged bats)*

The family Thyropteridae is composed of five species of small-sized aerial insectivorous bats, all belonging to the genus *Thyroptera*.

The common name, disc-winged bats, derives from the characteristic fleshy pads ('suckers') present at the base of the thumbs and ankles that are used to cling to the smooth walls of unfurling leaves of *Heliconia* and related banana-like plants in which they roost. As these leaves change from folded-up to flat, bats have to find another leaf with the proper shape in which to roost, so their small colonies are constantly on the move.

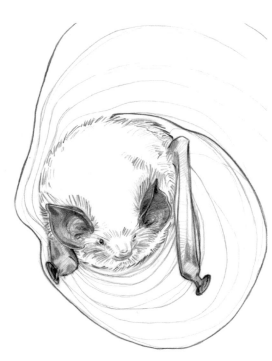

Disc-winged bats tend to inhabit moist tropical rainforests and are found from southern Mexico to southern Brazil. The family is regarded as primitive and its members have a domed skull, a slender muzzle and, like the Natalidae and Furipteridae, funnel-shaped ears.

Their small thumbs are also characteristic, and a short portion of the tail extends beyond the interfemoral membrane. The dorsal fur is brownish-to-black and some species have whitish ventral fur.

Thyroptera tricolor

Thyroptera tricolor

Thyropteridae *(disc-winged bats)*

Thyroptera (disc-winged bats)

1a. Thumb with oval disk; ventral fur bicoloured or tricoloured.

2a. FA > 35 mm; ventral fur bicoloured; forearm barely haired near the body.

3a. Ventral fur clearly frosted; hairs dark brown-to-blackish at the base, with pale-brown-to-whitish tips; calcar without lobes (sometimes just a single faintly developed lobe). *Thyroptera devivoi* (35–38 mm)

3b. Ventral fur bicoloured, not frosted; calcar with one well-developed lobe. *Thyroptera lavali* (37–41 mm)

2b. FA < 34.5 mm; ventral fur tricoloured; forearm densely hairy. *Thyroptera wynneae* (33–35 mm)

1b. Thumb with round disk; ventral fur unicoloured.

2a. Ventral fur white or pale grey; forearm slightly hairy near the body. *Thyroptera tricolor* (33–40 mm)

2b. Ventral fur yellowish-brown; forearm densely hairy. *Thyroptera discifera* (32–35 mm)

1a   1b   1a 2a **3a**   1a 2a **3b**

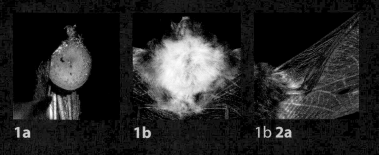

1a   **1b**   1b **2a**

Potential *Thyroptera* roost

Furipteridae *(smoky bats)*

The Furipteridae, known as smoky bats, is one of the smallest bat families and only contains two species: the smoky bat *Amorphochilus schnablii* and the thumbless bat *Furipterus horrens*. These small insectivorous bats have relatively long wings, domed skulls, funnel-shaped ears, and a delicate appearance, and resemble bats from the Thyropteridae and Natalidae to which they are closely related. The family's characteristic feature is the minute and functionless thumb, which is partly enveloped by the wing membrane. Its common name arises from the greyish colour of its fur. Of these two species, only the thumbless bat *Furipterus horrens* is known from the Amazon.

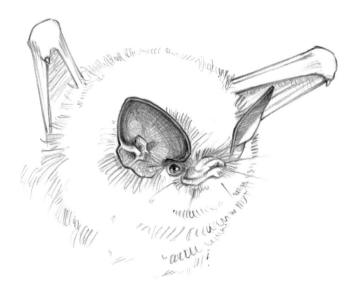

Furipterus (thumbless bat)

1a.   Only one species in the genus.

Furipterus horrens (30–40 mm)

Furipterus horrens

Noctilionidae (bulldog bats)

Commonly known as bulldog or fishing bats, the neotropical family Noctilionidae is represented by a single genus, *Noctilio*, containing two largely sympatric species, *Noctilio leporinus* and *Noctilio albiventris*. However, recent genetic evidence suggests that *N. albiventris* in fact consists of three lineages and that there is much cryptic diversity within this taxon.

Noctilionidae are found near water bodies from Mexico to Argentina (including the Caribbean Islands). They are medium-sized bats, with large drooping lips (hence the name 'bulldog bats') and relatively long legs. Their fur varies from orange to dark brown in colour and their wings are long and narrow.

The lesser bulldog bat *Noctilio albiventris* is mostly insectivorous, unlike the greater bulldog bat *Noctilio leporinus*. This latter species uses echolocation to detect ripples in water made by moving fish, which it then catches with its long legs and claws. Fish are eaten whilst perched and are sometimes stored in cheek pouches, an unusual feature in bats. This bat forages above coastal waters, rivers, and lakes, and can swim and even take off from the water surface.

Noctilio leporinus

Noctilio albiventris

Noctilionidae (bulldog bats)

Noctilio (bulldog bats)

1a. FA < 70 mm; feet and claws shorter than uropatagium.

Noctilio albiventris (54–70 mm)

1b. FA > 70 mm; feet and claws extend past the uropatagium.

Noctilio leporinus (70–90 mm)

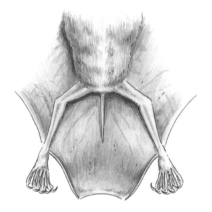

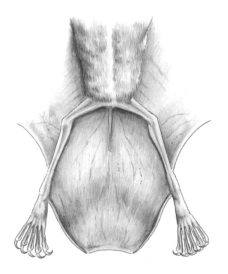

1a

1b

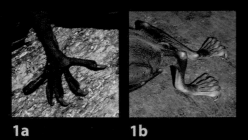

1a

1b

Amazon River

Mormoopidae *(moustached bats)*

The family Mormoopidae is composed of two genera, the moustached or naked-backed (*Pteronotus*) and the ghost-faced (*Mormoops*) bats. They are small to medium-sized and have characteristic wart-like projections above their nostrils and a small tail emerging from the dorsal surface of the uropatagium. Mormoopids are found from humid tropical to semiarid and arid sub-tropical habitats below 3,000 m throughout the New World, from the southwestern USA to southeastern Brazil, including the Greater Antilles.

The family's two genera can be separated by the presence of narrow (*Pleronotus*) or funnel-shaped (*Mormoops*) ears. The common names of *Pteronotus* are due to a peculiar fringe of long hairs around the mouth and to its wing membranes that, in some species, join over the middle of the back, giving an impression of hairlessness. The combination of hairs around the mouth and flaps on the lower lip are thought to funnel insects into the bat's mouth and focus echolocation pulses.

Pteronotus cf. *parnellii*

Pteronotus cf. *parnellii*

Mormoopidae *(moustached bats)*

Pteronotus (moustached bats)

1a. Bare back; wings attached on the middle of dorsum.
    2a. FA < 49 mm.

Pteronotus davyi (40–49 mm)

    2b. FA > 49 mm.

Pteronotus gymnonotus (50–55 mm)

1b. Furry back; wings attached to the side of body.
    2a. FA > 50 mm.

Pteronotus cf. *parnellii* * (50–63 mm)

    2b. FA < 50 mm.

Pteronotus personatus (40–48 mm)

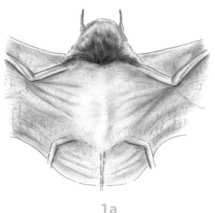 

1a   1b

---

* Cryptic species complex. We recommend using acoustic and/or genetic data for species identification (see also echolocation keys at the end of this guide, p. 112).

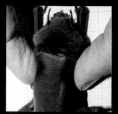
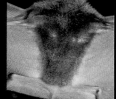

1a **1b**

Pteronotus cf. *parnellii*

Emballonuridae *(sac-winged bats)*

The Emballonuridae is a pantropical family that in the New World is found from northern Mexico to southern Brazil. Some neotropical species of this family possess sac-shaped glands near their shoulders, which explains the origin of the family's common name (sac-winged bats). These glands are usually more prominent in males and are used to produce pheromones.

Emballonurids are small aerial insectivorous bats, with relatively large eyes and long, narrow wings. These wings are so long that at rest they have one more fold than other bats. Most species are brown, but the four *Diclidurus* species, known as ghost bats, can vary from pale brownish to white and have distinctive pink wings, ears, and face. Most members of the genus *Saccopteryx* have two thin dorsal stripes that are especially evident in the greater sac-winged bat *S. bilineata*. Some, like the water-associated proboscis bat *Rhynchonycteris naso*, take advantage of their pale grey and yellowish fur to camouflage themselves on lichen-covered branches and wooden beams, and roost in a curious straight-lined, nose-to-tail formation.

Saccopteryx leptura

Centronycteris maximiliani

Emballonuridae *(sac-winged bats)*

1a. Fur white or whitish; wing sac on the uropatagium.

Diclidurus (p. 84)

1b. Fur not whitish; no wing sacs on the uropatagium.
    2a. Wing sacs absent.
        3a. Ears rounded; fur dark grey; small thumb.

Cyttarops (p. 84)

        3b. Combination not as above.
            4a. Muzzle long; forearm with several clusters
                of hair; fur grizzly brown; two pale stripes on
                back; wings attached to ankle.

Rhynchonycteris (p. 84)

            4b. Muzzle not long; forearm bare; fur yellowish
                or brownish; parallel rows of dots on the
                uropatagium; wings attached to base of toes.

Centronycteris (p. 84)

    2b. Wing sacs present in the propatagium.

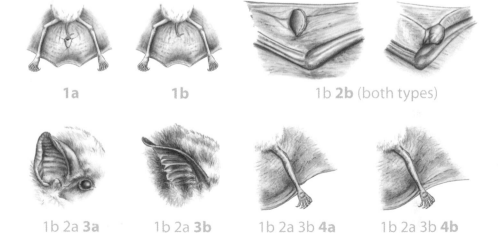

1a   **1b**   1b **2b** (both types)

1b 2a **3a**   1b 2a **3b**   1b 2a 3b **4a**   1b 2a 3b **4b**

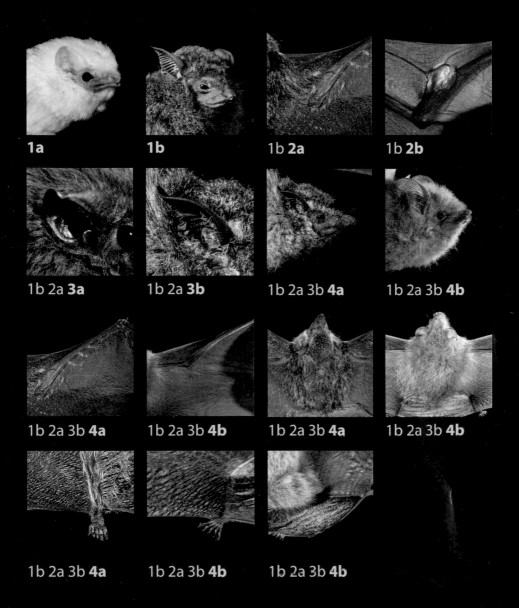

1a

1b

1b **2a**

1b **2b**

1b 2a **3a**

1b 2a **3b**

1b 2a 3b **4a**

1b 2a 3b **4b**

1b 2a 3b **4a**

1b 2a 3b **4b**

1b 2a 3b **4a**

1b 2a 3b **4b**

1b 2a 3b **4a**

1b 2a 3b **4b**

1b 2a 3b **4b**

Emballonuridae (sac-winged bats)

3a. Wing sac perpendicular to the forearm; fur on back without stripes.

    4a. Faint wing sacs not reaching (few mm) the edge of propatagium; wings attached near base of toe.

                                                Cormura (p. 84)

    4b. Wing sacs prominent, reaching the anterior edge of wing; wings attached above ankle.

                                                Peropteryx (p. 86)

3b. Wing sac close to the elbow and parallel to the forearm; two white lines on back (faint or absent in one species).

                                              Saccopteryx (p. 86)

3a **4a**   3a **4b**   3b

3a **4a**   3a **4b**

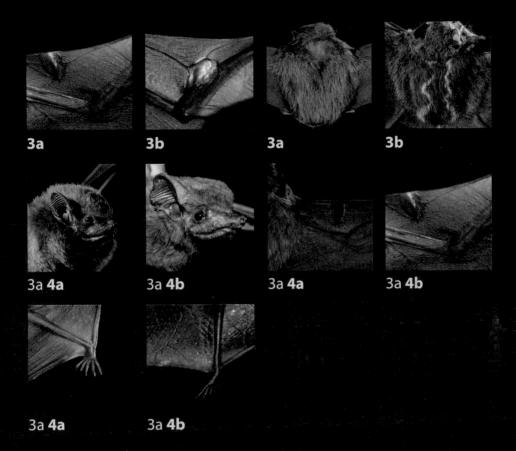

3a **3b** **3a** **3b**

3a **4a** 3a **4b** 3a **4a** 3a **4b**

3a **4a** 3a **4b**

Saccopteryx leptura

Emballonuridae *(sac-winged bats)*

Diclidurus (ghost bats)

1a. Wings pale brown; large thumb; fur sometimes dirty white.
Diclidurus isabellus (41–44 mm)

1b. Wings white or pale pink; small thumb; fur white.

    2a. FA > 69 mm.
Diclidurus ingens (70–73 mm)

    2b. FA < 69 mm.

        3a. FA > 60 mm.
Diclidurus albus (63–69 mm)

        3b. FA < 60 mm.
Diclidurus scutatus (51–59 mm)

Cyttarops (short-eared bat)

1a. Only one species in the genus.
Cyttarops alecto (45–47 mm)

Rhynchonycteris (proboscis bat)

1a. Only one species in the genus.
Rhynchonycteris naso (35–41 mm)

Centronycteris (shaggy bats)

1a. Body length < 65 mm.
Centronycteris maximiliani (41–45 mm)

1b. Body length > 65 mm.
Centronycteris centralis (42–48 mm)

Cormura (chestnut sac-winged bat)

1a. Only one species in the genus.
Cormura brevirostris (41–50 mm)

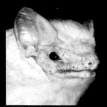

*Diclidurus
isabellus*

*Diclidurus
ingens*

*Diclidurus
albus*

*Diclidurus
scutatus*

*Cyttarops
alecto*

*Rhynchonycteris
naso*

*Centronycteris
centralis*

*Centronycteris
maximilani*

*Cormura
brevirostris*

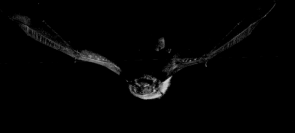

Rhynchonycteris naso

Emballonuridae (sac-winged bats)

Peropteryx (dog-like sac-winged bats)

1a. Wings white.

    2a.   Ears connected by band.

           Peropteryx leucoptera (41-46 mm)

    2b.   Ears not connected by band.

           Peropteryx pallidoptera (37–43 mm)

1b. Wings dark.

    2a.   FA > 43 mm.

           *Peropteryx kappleri / macrotis** (43–52 mm)

    2b.   FA < 43 mm.

           Peropteryx trinitatis (36–43 mm)

Saccopteryx (two-lined sac-winged bats)

1a. FA > 45 mm, dorsal fur black or dark.

           Saccopteryx bilineata (45–51 mm)

1b. FA < 43 mm, dorsal fur brown.

    2a.   Dorsal fur clearly bicoloured and frosted.

           Saccopteryx canescens (35–40 mm)

    2b.   Dorsal fur unicoloured or faintly bicoloured.

        3a.   FA < 36 mm; faint stripes on back; ventral fur unicoloured.

           Saccopteryx gymnura (33–36 mm)

        3b.   FA > 36 mm; distinct pale stripes on back; ventral fur bicoloured.

           Saccopteryx leptura (36–42 mm)

Saccopteryx

1a **2a**   1a **2b**

---

* We recommend classification as *P. kappleri / macrotis* until more external morphological data are available for reliable identification in the field.

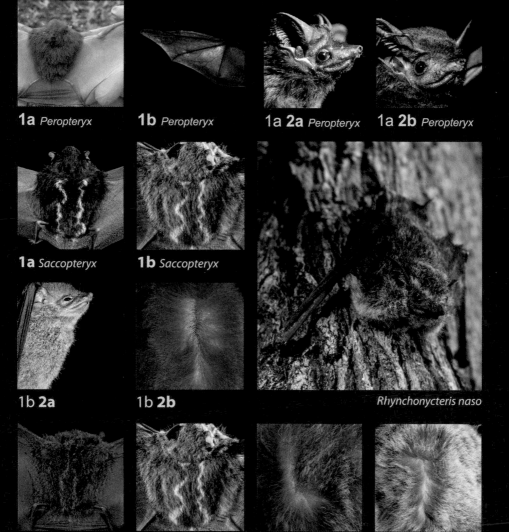

1a *Peropteryx*   **1b** *Peropteryx*   1a **2a** *Peropteryx*   1a **2b** *Peropteryx*

1a *Saccopteryx*   **1b** *Saccopteryx*

1b **2a**   1b **2b**   *Rhynchonycteris naso*

1b 2b **3a**   1b 2b **3b**   1b 2b **3a**   1b 2b **3b**

Vespertilionidae (evening bats)

Vespertilionids, commonly known as vesper or evening bats, are the largest bat family. This near-cosmopolitan family harbours more than 300 species and is present on all continents except Antarctica, as such being one of the most widespread of all mammalian groups. Five vesper bat genera are known from South America, all of which have been reported from the Amazon. They are small to large in size, have no noseleaf, and have ears with a simple tragus and usually large tail membranes that they use to capture the insects they prey upon. Vesper bats are mostly insectivorous but some Old World species have been reported to capture and consume fish and birds.

The genera *Eptesicus*, *Myotis* and *Rhogeessa* are mostly brown and black. However, the hairy-tailed bats of the genus *Lasiurus* are unusually colourful, and have long dense fur that can vary from bright yellow to red-orange. Another peculiarity of this genus is the extra pair of nipples (four in total) that allow females to give birth on occasions to quadruplets. They thrive in a wide range of habitats and exploit virtually all types of available roost sites.

Lasiurus sp.

Lasiurus egregius

Vespertilionidae *(evening bats)*

1a. Ears large 28–32 mm.

Histiotus (p. 92)

1b. Ears small < 27 mm.

  2a. Uropatagium furry.

Lasiurus (p. 92)

  2b. Uropatagium bare.

    3a. Two upper incisors.

Rhogeessa (p. 92)

    3b. Four upper incisors.

      4a. No gap after upper canine; tragus somewhat curved.

Eptesicus (p. 94)

      4b. Gap after upper canine; tragus straight and pointed.

Myotis (p. 96)

1a

1b

1b **2a**

1b **2b**

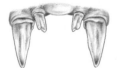

1b 2b **3a**

1b 2b **3b**

1b 2b 3b **4a**

1b 2b 3b **4b**

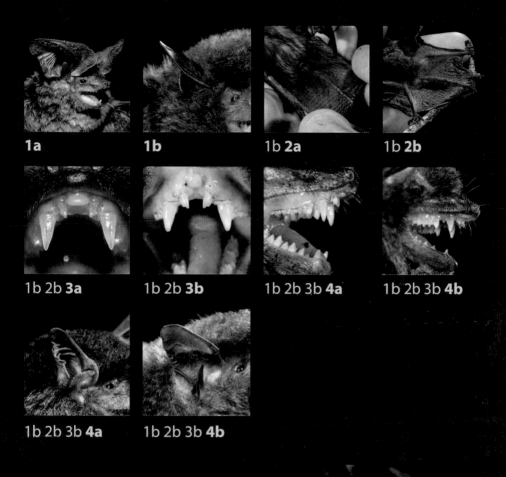

1a

1b

1b **2a**

1b **2b**

1b 2b **3a**

1b 2b **3b**

1b 2b 3b **4a**

1b 2b 3b **4b**

1b 2b 3b **4a**

1b 2b 3b **4b**

Myotis riparius

Vespertilionidae *(evening bats)*

Histiotus (evening big-eared bat)

1a. Only one species in the Amazon.

Histiotus velatus (42–50 mm)

Lasiurus (hoary bats)

1a. Dorsal fur reddish.

2a. Dorsal and ventral fur reddish.

Lasiurus egregius (48–50 mm)

2b. Dorsal fur reddish; venter brownish or greyish buff; wings reddish along the metacarpals.

Lasiurus blossevillii (36–43 mm)

2c. Dorsum reddish; venter blackish (sometimes with some white); wings completely black.

Lasiurus castaneus / atratus* (43–47 mm)

1b. Dorsal fur yellowish.

2a. Fur without frosting; FA < 51 mm.

Lasiurus ega (40–52 mm)

2b. Fur with frosted tips; FA > 50 mm.

Lasiurus cinereus (50–57 mm)

Rhogeessa (little yellow bats)

1a. FA < 30 mm.

Rhogeessa io (27–30 mm)

1b. FA > 29 mm.

Rhogeessa hussoni (29–31 mm)

---

* We recommend classification as *L. castaneus/atratus* until more external morphological data are available for reliable identification in the field.

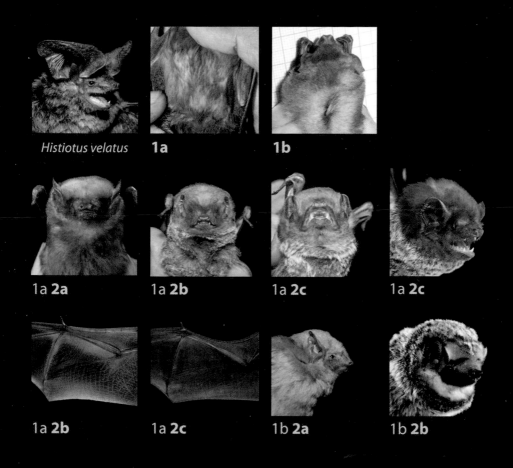

Histiotus velatus **1a** **1b**

1a **2a** 1a **2b** 1a **2c** 1a **2c**

1a **2b** 1a **2c** 1b **2a** 1b **2b**

Lasiurus egregius

Vespertilionidae *(evening bats)*

Eptesicus (big brown bats)

1a. Dorsal hair relatively short, < 7 mm.

    2a. FA < 37 mm.

                Eptesicus diminutus (30–37 mm)*

    2b. FA 36-43 mm; venter yellowish.

                Eptesicus furinalis (36–43 mm)*

    2c. FA > 41 mm; venter brownish.

                Eptesicus brasiliensis (40–47 mm)*

1b. Dorsal hair relatively long, > 7 mm.

    2a. FA < 44 mm.

                Eptesicus andinus (37–44 mm)*

    2b. FA > 43 mm.

                Eptesicus chiriquinus (42–49 mm)*

---

* In cases where measurements overlap, we recommend classification as *E. diminutus/furinalis*, *E. furinalis/brasiliensis* and *E. andinus/chiriquinus*, until more external morphological data are available for reliable identification in the field.

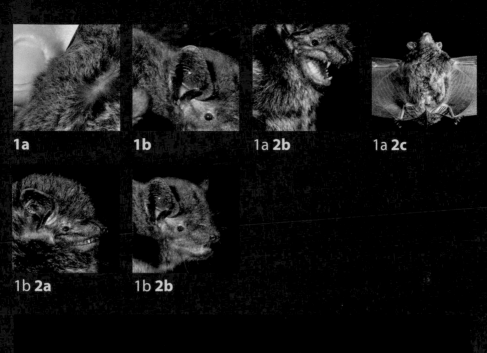

1a 1b 1a 2b 1a 2c

1b 2a 1b 2b

Eptesicus brasiliensis

Vespertilionidae *(evening bats)*

Myotis (little brown bats)

1a. Wings attached along the tibia.

Myotis simus (36–41 mm)

1b. Wings not attached along the tibia.

    2a. Dorsal fur very black and frosted; venter whitish.

Myotis albescens (31–37 mm)

    2b. Combination not as above.

        3a. Second upper premolar not aligned with other premolars.

Myotis riparius (31–38 mm)

        3b. Second upper premolar aligned with other premolars.

Myotis nigricans (30–38 mm)

1a   1b   1b 2b **3a**   1b 2b **3b**

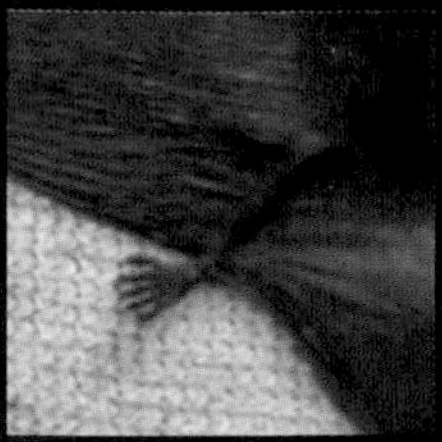

1a

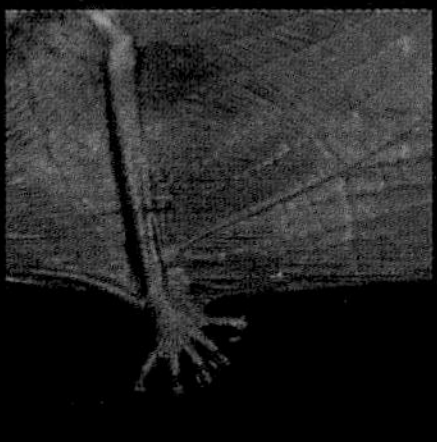

1b

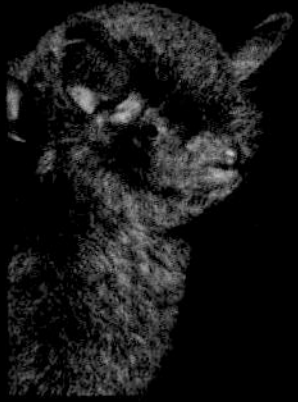

1b 2a

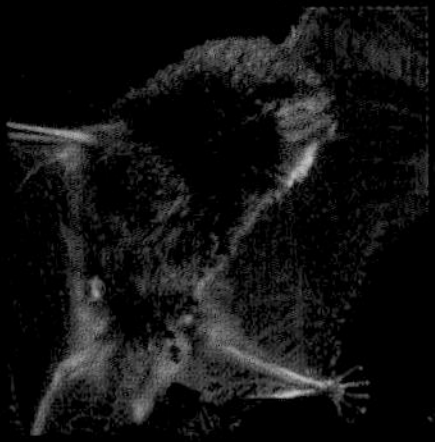

1b 2b

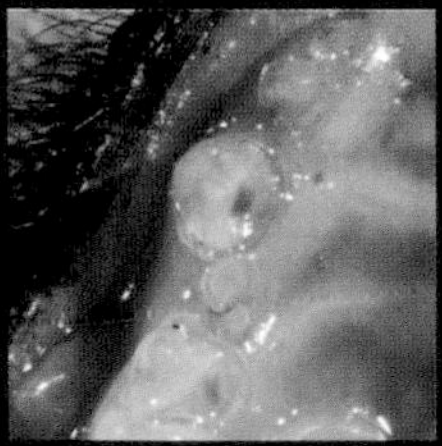

1b 2b 3a

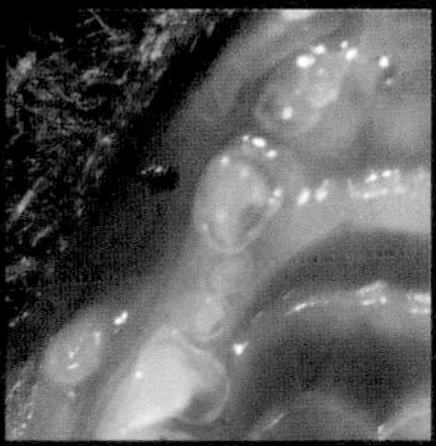

1b 2b 3b

Myotis riparius

Molossidae (free-tailed bats)

The Molossidae is a near-cosmopolitan family that, like the Vespertilionidae, is present on all continents. They are divided into two sub-families, the Molossinae and the Tomopeatinae, the latter including just one species, the blunt-eared bat *Tomopeas ravus* that is endemic to Peru. Molossids have relatively long, narrow wings and are adapted to rapid flight in open spaces. They are strong fliers and can cover large distances every night in search of food. Their common name, free-tailed bats, comes from their long tails that project beyond the uropatagium.

Their wing and tail membranes are usually very tough, their ears tend to be tilted forward, stiff and joined along part of their length, their legs are short and robust, and their feet have long sensory hairs. Neotropical species are mostly brown or black, although there are some exceptions such as the black mastiff bat *Molossus rufus*, which can be reddish in colour. Several species have throat glands that are less conspicuous in females.

Cynomops

Eumops auripendulus

Molossidae (free-tailed bats)

1a. Upper lip with deep vertical lines; ears joined.

Nyctinomops (p. 102)

1b. Upper lip with no vertical lines.

    2a. Muzzle between eyes and nose with a ridge.

        3a. Four lower incisors; lower posterior edge of ear thin and narrow.

Promops (p. 102)

        3b. Two lower incisors; lower posterior edge of ear flattened laterally.

Molossus (p. 104)

    2b. Muzzle flat, almost horizontal.

        3a. Squarish mouth when viewed ventrally, ears long and joined, reaching the nose when flattened.

Eumops (p. 106)

        3b. Triangular mouth when viewed ventrally; ears not joined, and do not reach the nose when flattened.

            4a. Muzzle and ears both pointed; dorsal fur continues onto face.

Molossops (p. 108)

            4b. Muzzle and ears both rounded; dorsal fur behind the ears.

Cynomops (p. 108)

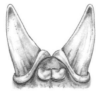

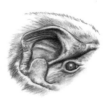

1b 2a **3a**   1b 2a **3b**   1b 2b 3b **4a**   1b 2b 3b **4b**

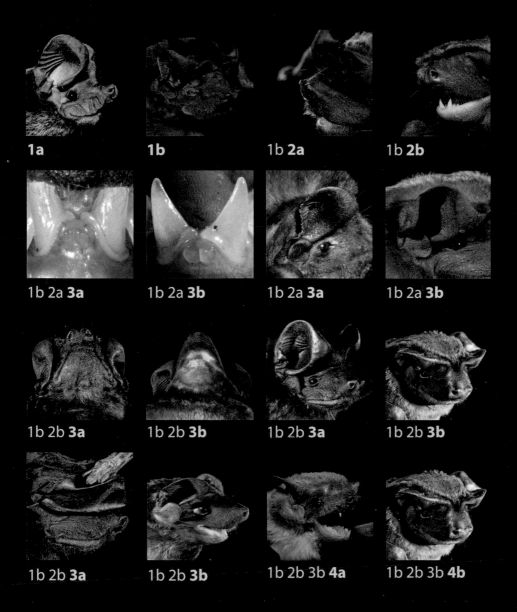

1a   1b   1b **2a**   1b **2b**

1b 2a **3a**   1b 2a **3b**   1b 2a **3a**   1b 2a **3b**

1b 2b **3a**   1b 2b **3b**   1b 2b **3a**   1b 2b **3b**

1b 2b **3a**   1b 2b **3b**   1b 2b 3b **4a**   1b 2b 3b **4b**

Molossidae (free-tailed bats)

Nyctinomops (broad-eared free-tailed bats)

1a. FA 58–65 mm.

Nyctinomops macrotis (58–65 mm)

1b. FA 48–53 mm.

Nyctinomops aurispinosus (48–53 mm)

1c. FA 40–48 mm.

Nyctinomops laticaudatus (40–48 mm)

Promops (crested mastiff bats)

1a. FA < 51 mm.

Promops nasutus (45–51 mm)

1b. FA > 51 mm.

Promops centralis (51–57 mm)

Temporary lake in the Amazon

Molossidae (free-tailed bats)

Molossus (common mastiff bats)

(There is great uncertainty regarding the taxonomy of this group.
Here we follow the nomenclature of Nogueira et al. 2014†)

1a. Dorsal fur unicoloured*

2a. FA 46–54 mm; face and membranes black (orangish-to-blackish fur).

*Molossus rufus** (46–54 mm)

2b. FA 41–49 mm; face and membranes not black, somewhat paler.

*Molossus pretiosus** (41–49 mm)

1b. Dorsal fur faint to clearly bicoloured.

2a. Dorsal fur faintly bicoloured.

3b. FA 35–37 mm.

Molossus coibensis (35–37 mm)

2b. Dorsal fur bicoloured.

3a. FA > 46 mm.

Molossus sinaloae (46–50 mm)

3b. FA 37–46 mm.

*Molossus molossus / currentium*** (37–46 mm)

3c. FA < 36 mm.

*Molossus sp.**** (33–36 mm)

---

* Highly variable fur colour. The photographs highlight some of the colour variation.
** We recommend classification as *M. molossus/currentium* until more external morphological data are available for reliable identification in the field.
*** Some records of smaller *Molossus* sp. exist that are awaiting phylogenetic and morphometric revision to determine their true taxonomic status.
† Nogueira, M.R., et al. 2014. Checklist of Brazilian bats, with comments on original records. *Check List*, 10(4), pp. 808–821.

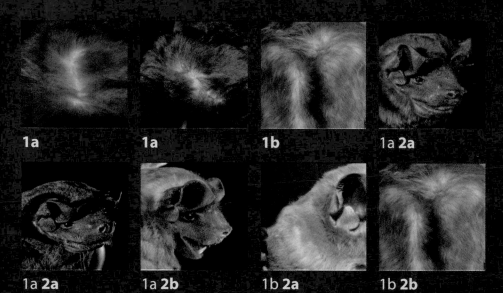

1a   1a   1b   1a 2a

1a 2a   1a 2b   1b 2a   1b 2b

Molossus rufus

Molossidae (free-tailed bats)

Eumops (bonnetted bats)

1a. FA < 55 mm.

    2a. Band of pure white fur along the venter/wing border.

        Eumops maurus (51–53 mm)

    2b. No band of pure white fur along the venter/wing border.

        3a. FA < 41 mm.

            Eumops hansae* (37–41 mm)

        3b. FA > 43 mm.

            Eumops bonariensis / delticus** (46–50 mm)

1b. FA > 55 mm.

    2a. Ear > 35 mm.

        3a. FA > 74 mm.

            Eumops perotis (75–84 mm)

        3b. FA < 74 mm.

            Eumops trumbulli (58–75 mm)

    2b. Ear < 35 mm.

        3a. Tragus pointed.

            Eumops auripendulus (54–68 mm)

        3b. Tragus broad and square.

            Eumops glaucinus (56–65 mm)

1b 2b **3a**   1b 2b **3b**

* Sometimes considered as a cryptic species complex including *E. nanus*.
** We recommend classification as *E. bonariensis / delticus* until more external morphological data are available for reliable identification in the field.

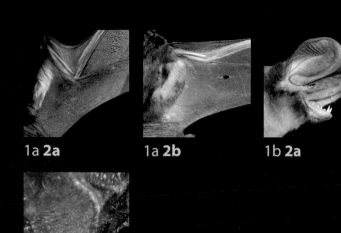

1a **2a**   1a **2b**   1b **2a**   1b **2b**

1b 2b **3a**

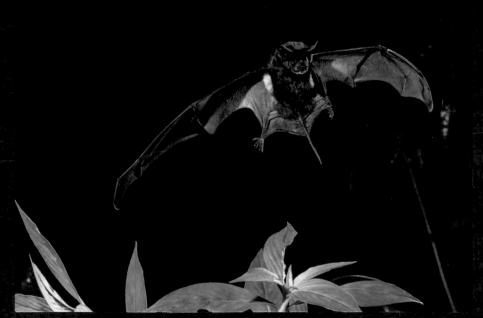

Eumops maurus

Molossops (dog-faced bats)

1a. FA with tiny bumps.

Neoplatymops mattogrossensis (27–33 mm)

1b. FA with no bumps.

    2a. FA > 34 mm; venter dark.

Molossops neglectus (34–37 mm)

    2b. FA < 33 mm; venter frosted.

Molossops temminckii (27–32 mm)

Cynomops (dog-like bats)

1a. FA > 40 mm.

Cynomops abrasus (40–52 mm)

1b. FA < 40 mm.

    2a. Four lower incisors; dorsal fur dark brown, venter pale.

Cynomops planirostris/paranus* (29–37 mm)

    2b. Four lower incisors; dorsal and ventral fur uniformly dark brown

Cynomops milleri (30–33 mm)

    2c. Two lower incisors; dorsal fur chestnut.

Cynomops greenhalli (33–39 mm)

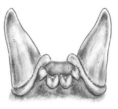 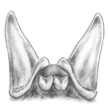

(Cynomops)

1b **2a**   1b **2b**

---

* We recommend classification as *C. planirostris / paranus* until more external morphological data are available for reliable identification in the field.

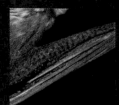

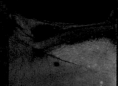

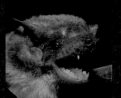

1a *Neoplatymops mattogrossensis*

1b *Molossops* sp.

1b 2a *Molossops neglectus*

1b 2b *Molossops temminckii*

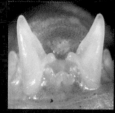

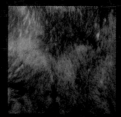

1b 2a *Cynomops*

1b 2a *Cynomops*

1b 2a *Cynomops*

1b 2c *Cynomops*

Cynomops abrasus

Natalidae (funnel-eared bats)

Natalus macrourus

This neotropical family comprises three genera containing six species of small, delicate, insectivorous bats. The family's common name, funnel-eared bats, derives from their large forward-pointing, funnel-like ears. These bats are characterised by their short thumbs and unusually long legs and tails. Their wings are broad, thereby giving good manoeuvrability that facilitates their gleaning foraging strategy. Funnel-eared bats roost colonially in humid caves.

Natalidae are distributed from Paraguay to northern Mexico and the West Indies, where they reach their greatest diversity. Only two species, *Natalus macrourus* and *N. tumidirostris*, are known to occur in South America and both have been recorded in the Amazon.

1a. Rostrum swollen.

Natalus tumidirostris (36–42 mm)

1b. Rostrum not swollen.

Natalus macrourus (35–41 mm)

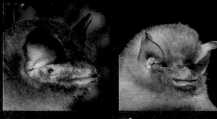

1a   1b

Maroaga Cave – Presidente Figueiredo, Amazonas State, Brazil

Echolocation

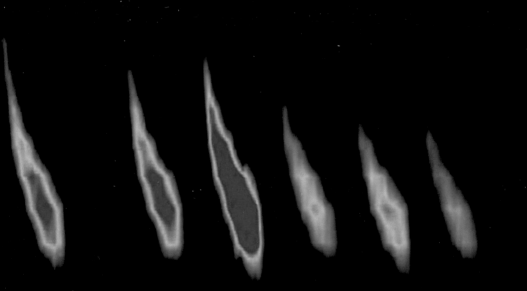

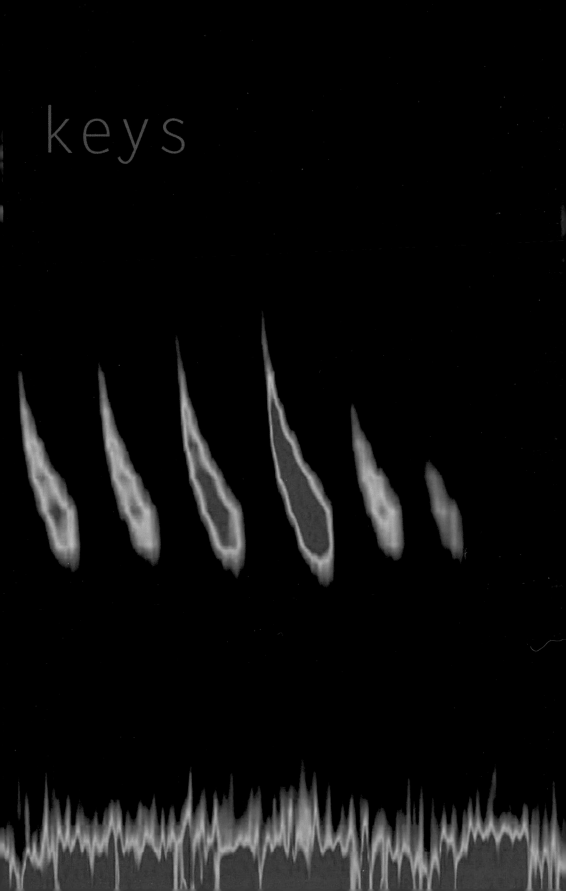

Echolocation keys

Across most of the neotropics, aerial insectivorous bats remain poorly studied. Aerial-hawking insectivorous bats are usually difficult to capture by mist-netting and the best technique for studying them is the use of ultrasound detectors. However, the echolocation calls of many neotropical aerial insectivorous bats are still inadequately described. Thus, intensified research efforts are urgently required to fill gaps in knowledge so that acoustic sampling can be used to its full potential in environmental impact assessments and monitoring programmes.

In terms of acoustic sampling techniques, the advent of automatic and fully autonomous recording stations has opened up new avenues for studying neotropical aerial insectivorous bats. However, reliable analysis of the data generated by acoustic surveys and monitoring studies requires the creation of a good call reference library for the bats of the study region. Currently, this kind of information is largely lacking for areas such as the Amazon.

It is well known that some species' echolocation calls are often similar and have considerable overlap in frequencies, which can complicate identification and even render findings unreliable. In addition, factors such as weather conditions, geographic location, habitat structure, flight height, and various other physiological and environmental factors can give rise to great variation in call structure within a particular species. Sex, age and reproductive status are other sources of variation, as has been found for several species. Thus, it is essential to quantify differences in echolocation call structure within and among tropical species to allow accurate acoustic assessments. It is also well known that handling and processing bats after capture can alter call properties due to the stress caused to individuals, and this is one of the main problems that arises when attempting to obtain high-quality recordings for reference libraries.

Rhynchonycteris naso

Several techniques such as discriminant function analysis, as well as, more recently, the use of synergetic pattern recognition algorithms in real time and artificial neural networks, have been employed in species identification based on echolocation call data. However, in order to develop and successfully use these techniques, an accurate description of the characteristics of the echolocation calls of all species known to occur in the study area is paramount. In the end, even with the development of new algorithms and techniques for automatic call identification, manual cross-checking and revision of results by experts remain essential.

Echolocation keys

Bat calls are highly variable due to numerous factors such as the type of activity and surrounding environmental clutter.

This variation often exacerbates overlap in the characteristics of the calls of certain species that can complicate the use of identification keys.

How should measurements be taken?

In order to use this key properly, it is essential to understand and standardize how measurements of calls are taken. All measurements must be taken from the harmonic that concentrates most energy, which, although varying from family to family, is usually the first or the second. All harmonics will be integer multiples of the "fundamental" frequency (first harmonic).

The frequency of maximum energy (FME) is extracted from the power spectrum as the frequency that is recorded at the moment of greatest call intensity.

Maximum and minimum frequencies can be measured on the power spectrum or on the spectrogram at the moment that the pulse differs most from the background noise. Thus, bandwidth should be calculated as the difference between the maximum and the minimum frequencies.

Start and end frequencies must be measured at the point where the amplitude of the oscillogram begins to consistently rise or decrease above the background noise. This can be obtained from the spectrogram when the intensity of the call is 20dB above the background noise. Accordingly, the call duration is measured between the start and the end point of the pulse.

Although not commonly referred to in other available keys, pulse intervals may be of interest and are defined as the time between the start of one pulse and the start of the subsequent one.

CF:	Constant frequency	**u:**	Upward modulated
QCF:	Quasi-constant frequency	**d:**	Downward modulated
FM:	Frequency modulated	**BW:**	Bandwidth
FME:	Frequency of maximum energy	**MinFreq:**	Minimum frequency
EF/SF:	End frequency / Start frequency	**MaxFreq:**	Maximum frequency

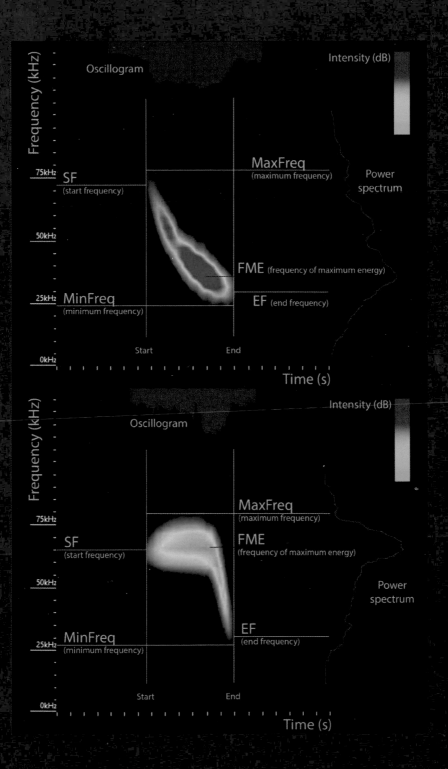

Echolocation keys

Some important issues to consider before deciding to work with echolocation data

Identification of neotropical bat species by their echolocation calls is a challenging task. As stated at the beginning of this key, calls are very plastic. Some species have distinctive calls that are easy to identify, while others substantially overlap with those of other taxa, thereby making reliable species identification difficult, if not impossible. It is thus essential that anyone aiming to analyse bat acoustic data has appropriate training to minimize data misinterpretation. This is true for both scientific studies and environmental impact assessments carried out by local consultants. Bat acoustic assessments heavily depend on the quality of the recordings since poor recordings can negatively affect identification success and the reliability of results. Thus, it is vital to understand not only how to analyse acoustic recordings but also how to properly set up detectors, calibrate microphones, and use specific recording settings (e.g. background filtering and frequency triggers).

Due to the rapid increase in the number of people using acoustics as a tool for surveying and monitoring bats, several automatic algorithms are now available that can speed up classification work. The positive aspect of these algorithms is that they can generate standardized results from massive datasets with little time commitment by the researcher. On the other hand, even though call analysis by experienced researchers is subjective and much more time consuming, manual call classification can give more accurate results in terms of identifying rare species, quantifying true diversity, and the presence of feeding buzzes and social calls, which are neglected in all available automatic identification software. The best processing method will clearly depend on the type of data that is hoped to be extracted from recordings and the objectives of the study. Remember that the amount of bat activity is fairly well correlated with the true number of bats flying in the area. However, bat activity is rarely comparable between species due to differences in the detectability of their calls and dissimilarities in the structure of their calls. In conclusion:

1. Understand, prepare and place correctly your equipment in the field (attend training sessions if necessary).
2. Store your data adequately (labelled, georeferenced, and including a description of the relevant metadata).
3. State the details of the specific detector settings that were used and calibrate the microphones.
4. If you aim to quantify relative abundance, specify how exactly you will quantify it.
5. Decide which species or species-group categories will be used to classify the recordings.
6. If you combine automatic and manual classifications, explain in detail how the manual verification is undertaken and the reasoning behind your choice of specific species-groups and the limitations of your analysis.
8. Understand the limitations of your equipment, take special care when analysing the data and exercise caution when interpreting your results.
9. Due to substantial variation in species detectability (e.g. quieter vs. louder calls), activity levels between species are rarely comparable.

Some notes on identification at family level

The following pages contain two acoustic keys, one for when harmonics are clearly recorded and the other for when they are not.

> If the harmonics cannot be distinguished in the sonograms, try to adjust the gain and filters in your analysis software in order to detect weaker harmonics and thus be able to use the first key (much simpler and more reliable). If you cannot find the harmonics, follow the second key step-by-step, but be very careful with confusing or faint pulses.

Do not worry about leaving many recordings as either "unidentified" or classified in "phonic groups" (including multiple species). This is preferable to ending up with a large number of incorrect species identifications.

> Misidentifications can lead to bad management decisions and therefore it is always better to rely on fewer but good-quality data rather than a massive amount of low-quality data.

Take into account the shape of pulses and the type of environment in which bats are recorded. Bats in highly cluttered habitats tend to greatly modulate their pulses. On the other hand, in open habitats calls tend to lose their modulated component and pulses may resemble emballonurid or molossid calls due to their almost constant-frequency components. The calls of the Molossidae and Vespertilionidae families are the most variable and can easily lead to misidentifications.

Echolocation keys

Main phonic-group selection
(if you *do* have harmonics recorded)

1a. FME located in the first harmonic.

    2a. Pulses with a short CF section and a long FMd tail (FME < 70kHz).

        Noctilionidae (p. 126)

    2b. Mostly QCF (at least in one of the pulse types, when call sequences include alternating pulse types); sometimes with small FM tails.

        Molossidae* (p. 134)

    2c. FM with final QCF part (very variable proportions of each type).

        Vespertilionidae – Thyropteridae* (p. 140)

1b. FME located in any other harmonic.

    2a. Pulses with at least one CF section.

        Mormoopidae (p. 126)

    2b. Mostly QCF, sometimes with small FMd tails.

        Emballonuridae (p. 128)

    2c. FM with final QCF section, FME > 110kHz.

        Natalidae (p. 138)

    2d. Only FM (extremely modulated pulses).

        3a. FME ≈ 130-170kHz.

        Furipteridae (p. 138)

---

* Be aware of the great variability found in this group.

Freq (kHz) Fundamental harmonics

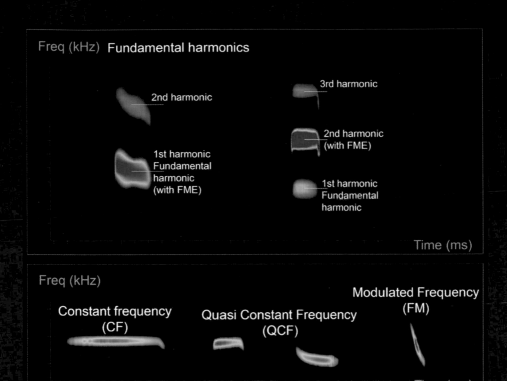

2nd harmonic

3rd harmonic

2nd harmonic
(with FME)

1st harmonic
Fundamental
harmonic
(with FME)

1st harmonic
Fundamental
harmonic

Time (ms)

Freq (kHz)

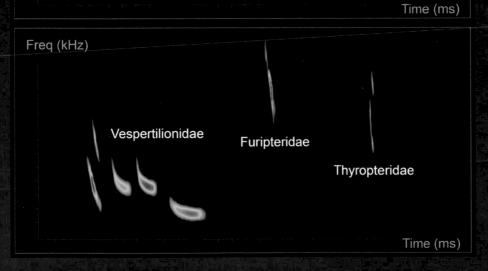

Constant frequency
(CF)

Quasi Constant Frequency
(QCF)

Modulated Frequency
(FM)

Time (ms)

Freq (kHz)

Vespertilionidae

Furipteridae

Thyropteridae

Time (ms)

Echolocation keys

Main phonic-group selection

(if you *do not* have harmonics recorded)

1a. Pulses with one CF section and a long FMd tail (FME < 70kHz).

Noctilionidae (p. 126)

1b. Pulses with at least one CF section.

Mormoopidae (p. 126)

1c. Mostly QCF (at least in one of the pulse types, when call sequences include alternating pulse types); sometimes with small FM tails.

    2a. QCF/ FMd (FME > 80kHz)

Emballonuridae A (p. 128)

    2b. Convex QCFu with FMd tails at the beginning of the pulses (1 or 2 types of pulses).

Emballonuridae B (p. 130)

    2c. Convex QCFd with FMd tails at the end of the pulses (1 or 2 types of pulses).

Emballonuridae C (p. 130)

    2d. Convex QCFd with a small FMu tail (two or three types of pulses).*

Molossidae A (p. 134)

    2e. Convex QCFu and concave QCFd.

Molossidae B (p. 136)

    2f. Convex QCFd and concave QCFd.

Molossidae C (p. 138)

    2g. Concave QCF (FME < 30kHz) (one or two types of pulses).

Molossidae D (p. 138)

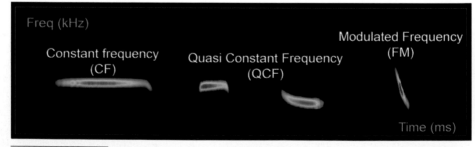

Freq (kHz)

Constant frequency (CF) Quasi Constant Frequency (QCF) Modulated Frequency (FM)

Time (ms)

---

* Be careful with the upper pulses, as they sometimes cannot be properly recorded, which can lead to misidentification. Third pulse sometimes concave QCFd.

Echolocation keys

1d. FM with final QCF (very variable proportions of each type).
     FME ≈ 30–100 kHz.

Vespertilionidae – Thyropteridae (p. 140)

1e. FM with final QCF with FME > 110 kHz.

Natalidae (p. 138)

1f. Only FM (extremely modulated pulses).
     2a.   FME ≈ 130–170 kHz.

Furipteridae (p. 138)

Detector hanging in the forest canopy

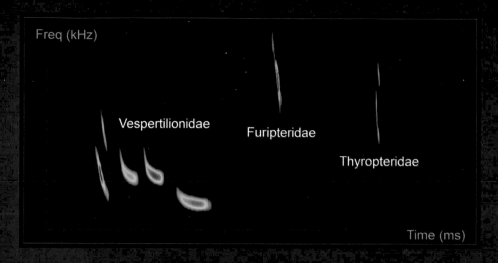

Freq (kHz)

Vespertilionidae   Furipteridae

Thyropteridae

Time (ms)

Cynomops planirostris

Echolocation keys

Noctilionidae

1a. CF / FMd, sometimes alternating with QCF.
   SF(CF) ≈ 68-76kHz

Noctilio albiventris

1b. CF / FMd, sometimes alternating with QCF.
   SF(CF) ≈ 53-61kHz

Noctilio leporinus

Some notes on the identification of Mormoopidae

Some genera of mormoopid bats can contain several cryptic species, and geographic variation may turn out to be greater in mormoopid bats than in other families. Specifically, *Pteronotus parnellii* seems to be a complex, comprising more than two sympatric species in the Amazon that can be easily separated by non-overlapping peak frequencies.

Mormoopidae

1a. CF / FMd (or small FMu / CF / FMd)

   2a. CF ≈ 55 kHz

Pteronotus cf. parnellii 55 kHz

   2b. CF ≈ 60 kHz

Pteronotus cf. parnellii 60 kHz

1b. CF / FMd / CF; SF(CF) ≈ 55 kHz

Pteronotus gymnonotus

1c. CF / FMd / CF; SF(CF) ≈ 68–69 kHz

Pteronotus personatus

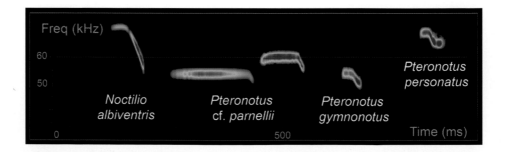

Freq (kHz) *Pteronotus* cf. *parnellii*

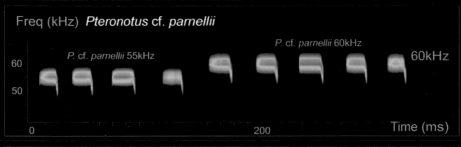

Freq (kHz) *Pteronotus* cf. *parnellii* - Feeding buzz

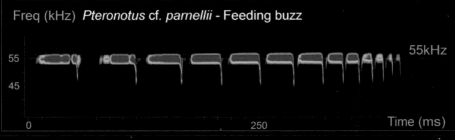

Freq (kHz)

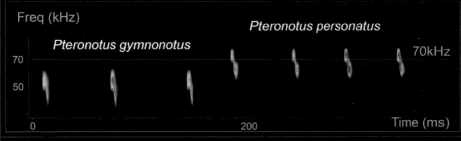

Freq (kHz) *Pteronotus personatus* - Feeding buzz

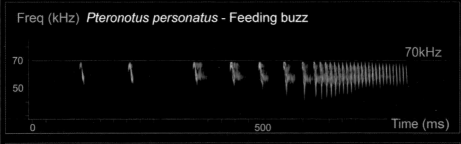

Freq (kHz) *Noctilio albiventris*

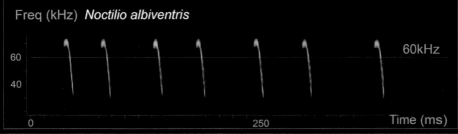

Echolocation keys

Some notes on the identification of Emballonuridae

One of the most useful features for separating emballonurid species and phonic groups is the alternation of different call frequency types. However, this can be a source of misidentification. The problem lies in the fact that the last upper pulse is sometimes not recorded due to its low intensity or simply because some bats might not emit it under certain conditions. It is thus recommended to adjust the gain to try to highlight these faint pulses. If one fails to take this into account, the activity of the genus *Centronycteris* or of species such as *Saccopteryx gymnura/canescens* could be greatly overestimated, whereas that of *Saccopteryx leptura* or *Saccopteryx bilineata* could be underestimated.

Another point to bear in mind is how to determine the slope angle when separating the groups *Centronycteris/Saccopteryx* from *Diclidurus/Peropteryx* spp. Low-quality recordings with a lot of confusing background noise and faint calls are common and to avoid this it is sometimes a good idea to switch your full spectrum sonograms to a zero-crossing representation to improve the detection of the angle of the pulses.

Emballonuridae

1a. QCF/ FMd; QCF ≈ 100 kHz.

Emballonuridae A

1b. Convex QCFu with FMd tails at the beginning of the pulses.

Emballonuridae B

1c. Convex QCFd with FMd tails at the end of the pulses.

Emballonuridae C

Emballonuridae A

1a. Only one species with this type of pulse.

Rhynchonycteris naso

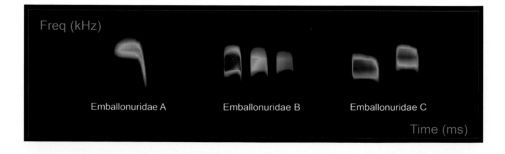

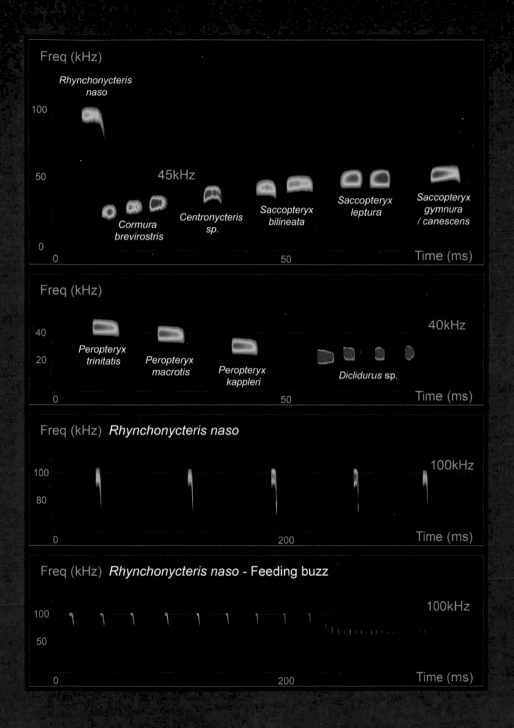

Freq (kHz)

Rhynchonycteris naso

100

50

45kHz

Cormura brevirostris

Centronycteris sp.

Saccopteryx bilineata

Saccopteryx leptura

Saccopteryx gymnura / canescens

0

0   50   Time (ms)

Freq (kHz)

40

20

40kHz

Peropteryx trinitatis

Peropteryx macrotis

Peropteryx kappleri

Diclidurus sp.

0   50   Time (ms)

Freq (kHz) *Rhynchonycteris naso*

100

80

100kHz

0   200   Time (ms)

Freq (kHz) *Rhynchonycteris naso* - Feeding buzz

100

50

100kHz

0   200   Time (ms)

Echolocation keys

Emballonuridae B

1a. One single pulse type.

    2a. FME ≈ 54 kHz.

Emballonuridae I
(Saccopteryx gymnura / canescens)

    2b. FME ≈ 40 kHz.

Emballonuridae II
(Centronycteris centralis / maximiliani)

    2c. FME ≈ 35 kHz.

Cyttarops alecto

1b. Two alternating types of pulses.

    2a. Lower pulse FME ≈ 48 kHz.
        Higher pulse FME ≈ 55 kHz.

Saccopteryx leptura

    2b. Lower pulse FME ≈ 42 kHz.
        Higher pulse FME ≈ 45 kHz.

Saccopteryx bilineata

1c. Three alternating types of pulses.

    2a. Lower pulse FME ≈ 25 kHz.
        Intermediate pulse FME ≈ 28 kHz.
        Higher pulse FME ≈ 30 kHz.

Cormura brevirostris

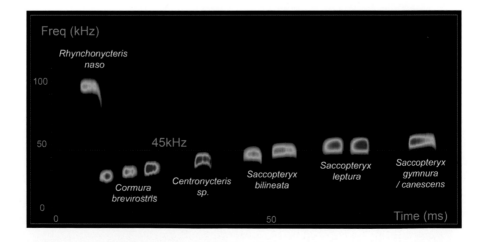

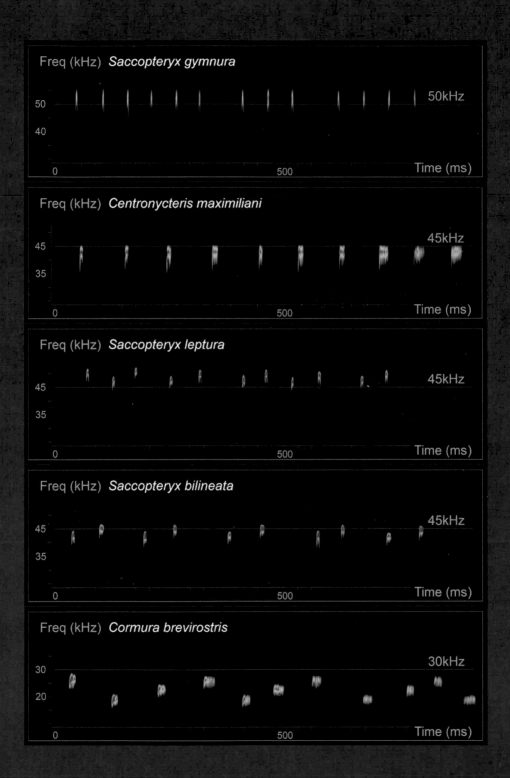

Echolocation keys

Emballonuridae C

1a. One type of pulse.

    2a. FME ≈ 42–44 kHz.

                                Peropteryx trinitatis

    2b. FME ≈ 37–39 kHz.

                                Peropteryx macrotis

    2c. FME ≈ 29–33 kHz.

                                Peropteryx kappleri

1b. Two alternating types of pulses*

    2a. Lower pulse FME ≈ 26 kHz
         Higher pulse FME ≈ 30 kHz

                                Diclidurus albus / scutatus

    2b. Lower pulse FME ≈ 19 kHz
         Higher pulse FME ≈ 22 kHz

                                Diclidurus ingens

How to separate *Diclidurus* and *Peropteryx* from molossid calls

Identification of species emitting low-frequency calls is challenging as calls are highly variable even within a single sequence. Due to the great overlap between the calls of some emballonurids (*Diclidurus* and *Peropteryx*) and molossid bats it is sometimes difficult to separate them.

We suggest following these steps:

1. Try to find the fundamental harmonic by adjusting the gain. If successful, genus separation is straightforward and clear.
2. Try to identify an obvious downturn at the end of the pulses, which is different from those in emballonurid species.
3. If it is impossible to see any harmonic, check the shape, angle and alternation.
4. If the calls overlap or show no clear patterns, it is recommended to classify them as "unidentified" which is the most conservative way of processing your data.
5. If you are not completely certain about an identification, consult a more experienced specialist.

---

* These groups can sometimes overlap. Then we recommend classification as *Diclidurus* spp.

Freq (kHz)

40

20 *Peropteryx trinitatis* *Peropteryx macrotis* *Peropteryx kappleri* *Diclidurus* sp.

40kHz

0 50 Time (ms)

Freq (kHz) *Peropteryx macrotis*

40 40kHz

30

0 600 Time (ms)

Freq (kHz) *Peropteryx kappleri* - Feeding buzz

40 30kHz

20

0 1000 Time (ms)

Freq (kHz) Fundamental harmonics

2nd harmonic

3rd harmonic

1st harmonic
Fundamental
harmonic
(with FME)

2nd harmonic
(with FME)

1st harmonic
Fundamental
harmonic

Time (ms)

Echolocation keys

Molossidae

1b. Convex QCFd with one initial FMu (three types of pulses).*

Molossidae A (p. 134)

1c. Convex QCFu and concave QCFd.

Molossidae B (p. 136)

1d. Convex QCFd and concave QCFd.

Molossidae C (p. 138)

1e. Concave QCF (FME < 30kHz).

Molossidae D (p. 138)

Molossidae A

1a. Lower pulse FME ≈ 33–35 kHz.
Intermediate pulse FME ≈ 35–40 kHz.
Higher pulse FME ≈ 40–45 kHz.

Molossus I*
Molossus molossus

1b. Lower pulse FME ≈ 25–30 kHz.
Intermediate pulse FME ≈ 30–35 kHz.
Higher pulse FME ≈ 35–40 kHz.

Molossus II*
Molossus sinaloae / currentium / rufus

---

* Be careful with the second and third upper pulses, as they sometimes cannot be properly recorded due to their low intensity, which can lead to misidentification. The first FMu part might not be present if the pulse is too faint. *Molossus* I & II can sometimes overlap. In some cases the higher pulse can be strongly modulated and may be followed by sequences of several similar modulated concave pulses (see figure).

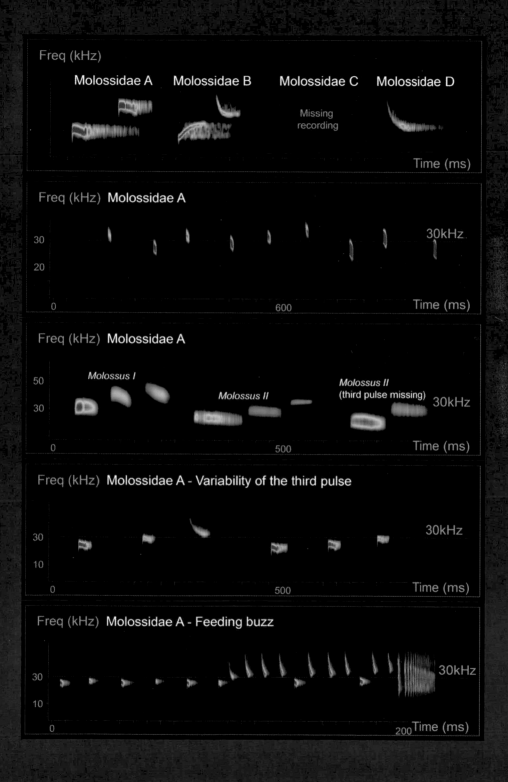

Echolocation keys

Molossidae B

1a.   Lower pulse < 40kHz.

    2a.   Lower pulse, EF ≈ 34 kHz.*
        Higher pulse, EF ≈ 37 kHz. *

Promops nasutus

    2a.   Lower pulse, EF ≈ 28 kHz.*
        Higher pulse, EF ≈ 30 kHz.*

Promops centralis

1b.   Lower pulse > 40kHz.

    2a.   Lower pulse, EF ≈ 54 kHz.*
        Higher pulse, EF ≈ 55 kHz.*

Molossops temminckii

    2a.   Lower pulse, EF ≈ 44 kHz.*
        Higher pulse, EF ≈ 46 kHz.*

Molossops neglectus

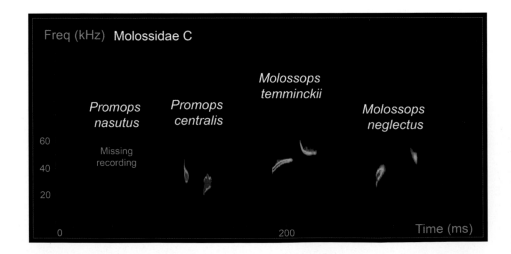

* These groups can sometimes overlap. Therefore, we recommend classification as *Molossus* spp., *Promops* spp. or *Molossops* spp. Be careful with the upper pulses, as they sometimes cannot be properly recorded, which can lead to misidentification.

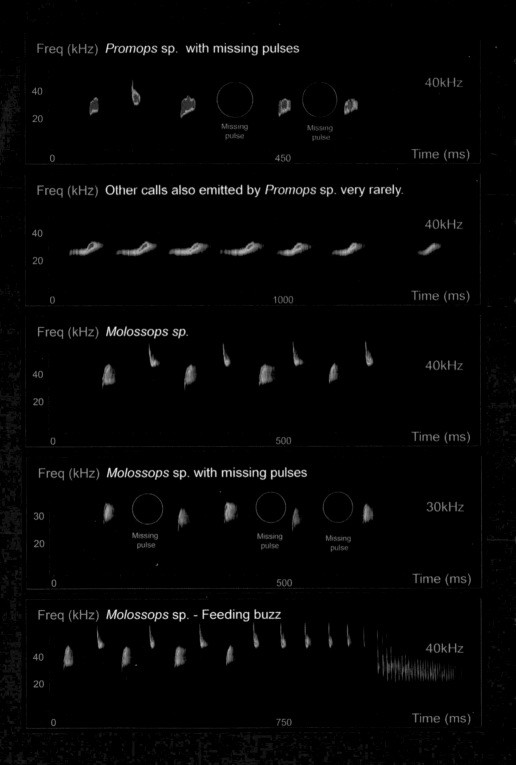

Freq (kHz) *Promops* sp. with missing pulses

40

20

Missing
pulse

Missing
pulse

40kHz

0 450 Time (ms)

Freq (kHz) Other calls also emitted by *Promops* sp. very rarely.

40

20

40kHz

0 1000 Time (ms)

Freq (kHz) *Molossops sp.*

40

20

40kHz

0 500 Time (ms)

Freq (kHz) *Molossops* sp. with missing pulses

30

20

Missing
pulse

Missing
pulse

Missing
pulse

30kHz

0 500 Time (ms)

Freq (kHz) *Molossops* sp. - Feeding buzz

40

20

40kHz

0 750 Time (ms)

Echolocation keys

Molossidae C

1a. Only one species with this type of pulse.

Neoplatymops mattogrossensis

Molossidae D

1a. Only one type of pulse.

Nyctinomops macrotis

1b. Two alternating types of two pulses.

2a. Lower pulse, EF ≈ 18 kHz.
Higher pulse, EF ≈ 22 kHz.

Molossidae III
Eumops auripendulus / glaucinus / dabbenei / hansae* / maurus
Nyctinomops laticaudatus, Tadarida brasiliensis
Cynomops planirostris / paranus / greenhalli / abrasus

Natalidae

1a. Only one type of pulse.

Natalus sp.

Furipteridae

1a. Only one species with this type of pulse.**

Furipterus horrens

Thyropteridae

1a. Only one genus with this type of pulse.***

Thyroptera sp.

---

* Sometimes considered as a cryptic species complex with *E. nanus*.
** Sometimes pulses seem to be grouped in sequences of 5–20 pulses during the search phase.
*** Sometimes confused with *Myotis riparius*. Therefore, we recommend classification as *M. riparius/ Thyroptera* spp.

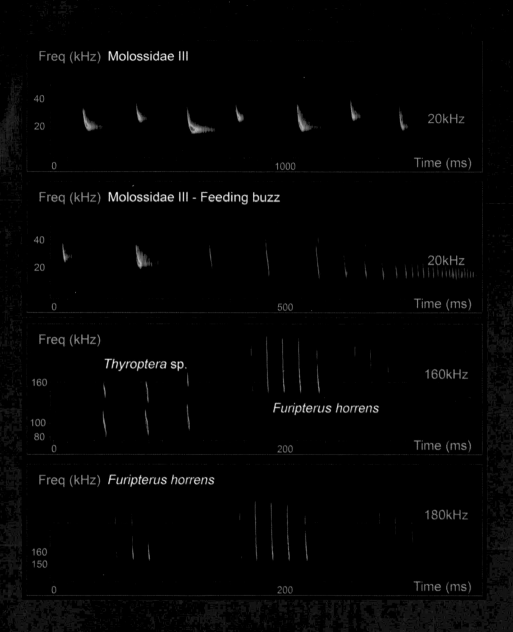

Freq (kHz) Molossidae III

40

20 — 20kHz

0 — 1000 — Time (ms)

Freq (kHz) Molossidae III - Feeding buzz

40

20 — 20kHz

0 — 500 — Time (ms)

Freq (kHz)

Thyroptera sp. — 160kHz

160

Furipterus horrens

100
80

0 — 200 — Time (ms)

Freq (kHz) *Furipterus horrens*

180kHz

160
150

0 — 200 — Time (ms)

Echolocation keys

Vespertilionidae

1a. Pulse mainly FMd; EF 25–45 kHz with irregular and
alternating sequences.*

   2a. EF ≈ 25–35 kHz.

   Vespertilionidae I
Lasiurus ega / castaneus / egregius / atratus

   2b. EF ≈ 40–45 kHz.

   Vespertilionidae II
Rhogeessa io / Lasiurus blossevillii

1b. Pulse initially FM, but with a considerable QCFd part.
Generally regular low frequencies.

   2a. EF ≈ 25–39 kHz.*

   Eptesicus I
Eptesicus brasiliensis / furinalis / chiriquinus

   2b. EF > 45 kHz; pulses ending with a QCF tail.

   3a. EF > 55 kHz.

   Myotis riparius

   3b. EF ≈ 45–50 kHz.

   Myotis nigricans

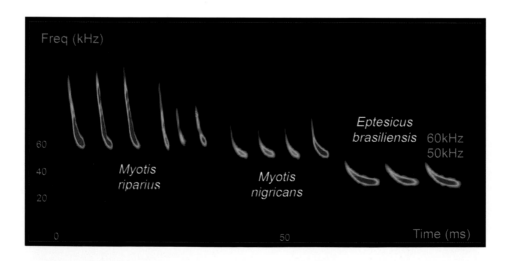

---

* These species can sometimes overlap in frequencies which may lead to misidentifications.

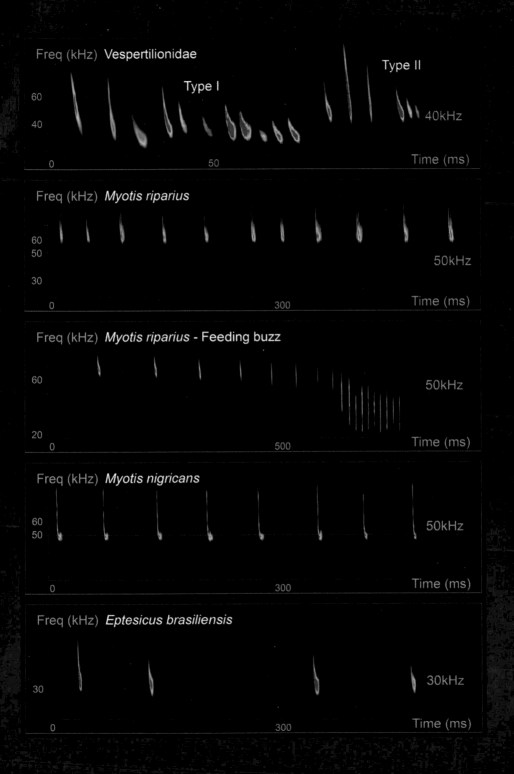

Appendix: Species rostra

Phyllostomidae sf. Desmodontinae

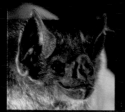
Desmodus rotundus

Diaemus youngi

Diphylla ecaudata

Phyllostomidae sf. Glossophaginae

Anoura caudifer

Anoura geoffroyi

Choeroniscus godmani

Choeroniscus minor

Glossophaga commissarisi

Glossophaga longirostris

Glossophaga soricina

Lonchophylla thomasi

Lichonycteris obscura

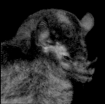
Lionycteris spurrelli

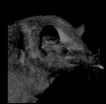
Scleronycteris ega

Phyllostomidae sf. *Stenodermatinae*

Ametrida
centurio

Artibeus
amplus

Artibeus
concolor

Artibeus
lituratus

Artibeus
obscurus

Artibeus
planirostris

Chiroderma
trinitatum

Chiroderma
villosum

Dermanura
anderseni

Dermanura
cinerea

Dermanura
glauca

Dermanura
gnoma

Enchisthenes
hartii

Mesophylla
macconnelli

Yet-to-be-
photographed

Platyrrhinus
aurarius

Platyrrhinus
brachycephalus

Platyrrhinus
fusciventris

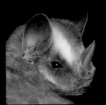

Platyrrhinus
incarum

Platyrrhinus
infuscus

Platyrrhinus
lineatus

Appendix: *Species rostra*

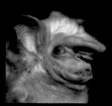

Sphaeronycteris toxophyllum (♂)

Sphaeronycteris toxophyllum (♀)

Sturnira lilium

Sturnira magna

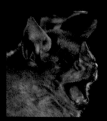

Sturnira tildae

Uroderma bilobatum

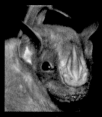

Uroderma magnirostrum

Vampyriscus bidens

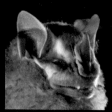

Vampyriscus brocki

Yet-to-be-photographed

Vampyressa melissa

Vampyressa pusilla/thyone

Vampyrodes caraccioli

Phyllostomidae sf. *Phyllostominae*

Chrotopterus
auritus

Glyphonycteris
daviesi

Glyphonycteris
sylvestris

Lampronycteris
brachyotis

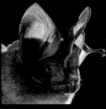

Lonchorhina
aurita

Lonchorhina
inusitata

Lophostoma
brasiliense

Lophostoma
carrikeri

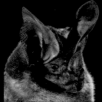

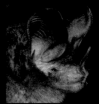

Lophostoma
schulzi

Lophostoma
silvicolum

Macrophyllum
macrophyllum

Micronycteris
brosseti

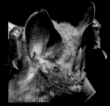

Yet-to-be-
photographed

Micronycteris
hirsuta

Micronycteris
homezorum

Micronycteris
megalotis

Micronycteris
microtis

Micronycteris
minuta

Micronycteris
sanborni

Micronycteris
schmidtorum

Appendix: Species rostra

*Mimon
bennettii*

*Mimon
crenulatum*

Yet-to-be-
photographed

*Neonycteris
pusilla*

*Phylloderma
stenops*

*Phyllostomus
discolor*

*Phyllostomus
elongatus*

*Phyllostomus
hastatus*

*Phyllostomus
latifolius*

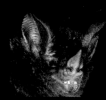

*Tonatia
bidens*

*Tonatia
saurophila*

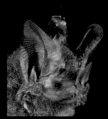

*Trachops
cirrhosus*

*Trinycteris
nicefori*

*Vampyrum
spectrum*

Phyllostomidae sf. *Carollinae*

*Carollia
benkeithi*

*Carollia
brevicauda*

*Carollia
castanea*

*Carollia
perspicillata*

*Rhinophylla
fischerae*

*Rhinophylla
pumilio*

Thyropteridae

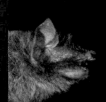

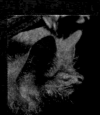

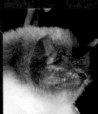

*Thyroptera
devivoi*

*Thyroptera
discifera*

*Thyroptera
lavali*

*Thyroptera
tricolor*

Furipteridae *Noctilionidae*

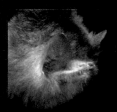

*Thyroptera
wynneae*

*Furipterus
horrens*

*Noctilio
albiventris*

*Noctilio
leporinus*

Appendix: Species rostra

Mormoopidae

*Pteronotus
davyi*

*Pteronotus
gymnonotus*

*Pteronotus
parnellii 55*

*Pteronotus
parnellii 60*

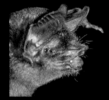

*Pteronotus
personatus*

Emballonuridae

*Centronycteris
centralis*

*Centronycteris
maximiliani*

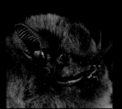

*Cormura
brevirostris*

*Cyttarops
alecto*

*Diclidurus
albus*

*Diclidurus
ingens*

*Diclidurus
isabellus*

*Diclidurus
scutatus*

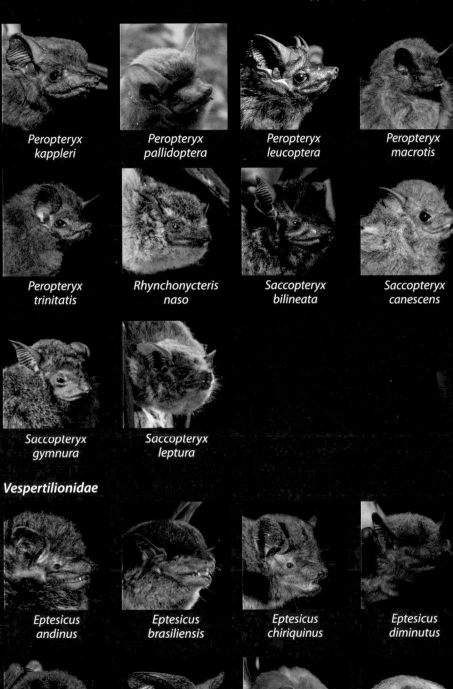

Peropteryx
kappleri

Peropteryx
pallidoptera

Peropteryx
leucoptera

Peropteryx
macrotis

Peropteryx
trinitatis

Rhynchonycteris
naso

Saccopteryx
bilineata

Saccopteryx
canescens

Saccopteryx
gymnura

Saccopteryx
leptura

Vespertilionidae

Eptesicus
andinus

Eptesicus
brasiliensis

Eptesicus
chiriquinus

Eptesicus
diminutus

Eptesicus
furinalis

Histiotus
velatus

Lasiurus
atratus

Lasiurus
blossevillii

Lasiurus castaneus

Lasiurus cinereus

Lasiurus ega

Lasiurus egregius

Myotis albescens

Myotis nigricans

Myotis riparius

Myotis simus

Rhogeessa hussoni

Rhogeessa io

Molossidae

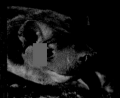

Cynomops
abrasus

Cynomops
greenhalli

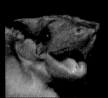

Cynomops
paranus

Cynomops
milleri

Cynomops
planirostris

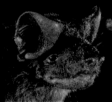

Eumops
auripendulus

Eumops
bonariensis

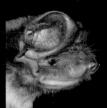

Eumops
glaucinus

Eumops
hansae

Eumops
maurus

Eumops
trumbulli

Eumops
perotis

Molossops
neglectus

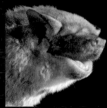

Molossops
temminckii

Molossus
rufus

Molossus
coibensis

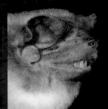

Molossus
currentium

Molossus
molossus

Molossus
pretiosus

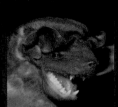

Molossus
sinaloae

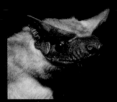

Neoplatymops mattogrossensis

Nyctinomops laticaudatus

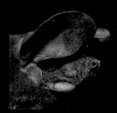

Nyctinomops macrotis

Nyctinomops aurispinosus

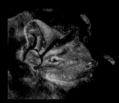

Promops centralis

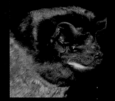

Promops nasutus

Natalidae

Natalus macrourus

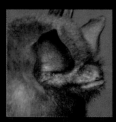

Natalus tumidirostris

Megascops watsonii, a predator of bats in the Amazon

Adrià López-Baucells

Adrià started working with the Bat Research Group, Granollers Museum of Natural Sciences (Catalonia), in 2005. Since then he has collaborated on several projects in fields such as habitat selection, biogeography, behaviour, and migration. He finished his BSc in Biology at the University of Barcelona in 2010 with a final project on neotropical bats based on fieldwork undertaken in Colombia. His MSc thesis was carried out in Sydney (Australia) on behavioural ecology and physiology in megachiroptera.

Currently, he is a PhD student under the supervision of Dr. Christoph Meyer, and Prof. Jorge Palmeirim. His PhD project uses autonomous ultrasound recording stations as a means of investigating the long-term impact of forest fragmentation on aerial insectivorous bats.

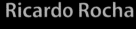

Ricardo Rocha

BSc in Biology by the University of Lisbon and MSc in Conservation Science by Imperial College London with thesis dedicated to São Tomé endemic birds' response to agricultural intensification. Following his MSc, he worked on the ecology of seabirds and endemic reptiles of the Selvagens archipelago (Portugal) and then moved to the Metapopulation Research Centre (Finland) to investigate the efficiency of Malagasy protected areas in reducing deforestation.

He has since worked with bird and bat ecology in Madagascar, Kenya and Brazil. His PhD, based at the Universities of Lisbon and Helsinki and supervised by Christoph Meyer, Jorge Palmeirim and Mar Cabeza, focused on the effects of tropical forest fragmentation on the spatio-temporal dynamics of phyllostomid bat communities. Ricardo is now based at the Department of Zoology, University of Cambridge.

Oriol Massana Valeriano

Oriol studied Biological Sciences at the University of Barcelona but then started working as an illustrator of educational, scientific, and cultural books and magazines. In 2010 he began to focus on 3D design and animation, which led to a career in nature documentaries. Another facet of his professional work focuses on photography, an interest that has been increasing since 2004. He has contributed to a number of publications and has received a number of awards.

Since 2011 he has worked with the Bat Research Group at the Granollers Museum of Natural Sciences (Catalonia) as a photographer, while developing a popular exposition on Catalan bats. During 2014 he spent three months in the Amazon rainforest as a volunteer on the bat research projects led by the other authors of this guide.

Authors & Illustrators

Paulo E. D. Bobrowiec

Bat researcher at the National Institute for Amazonian Research (INPA). He undertook a BSc in Biological Science at the Universidade Federal de Uberlândia and both a MSc and PhD at INPA. His main area of expertise is bat ecology and community structure, feeding strategies, and bat pollination and seed dispersal. His MSc was focused on the effect of secondary vegetation on bat communities, while his PhD was devoted to the feeding ecology of *D. rotundus*.

Enrico Bernard

Bat researcher and Professor of Conservation Biology at Universidade Federal de Pernambuco, Brazil.

He carried out his BSc in Biology in the Universidade de São Paulo in Ribeirão Preto. His MSc was conducted in Ecology at INPA (Manaus), where he worked on the vertical stratification of bat communities. His PhD was obtained from York University, Toronto, Canada, and focused on the effects of forest fragmentation on Amazonian bats.

Jorge Palmeirim

Associate Professor at the University of Lisbon and member of its cE3c research center, where he coordinates the Conservation Ecology Group. He is also a member of the IUCN Species Survival Commission (Chiroptera Specialist Group). He studied Biology at the Universities of Luanda and Lisbon, followed by an MA and PhD in Systematics and Ecology at the University of Kansas. His main research interests are the ecology and conservation biology of bats and birds, and tropical ecology.

Christoph F. J. Meyer

Lecturer in Global Ecology and Conservation at the University of Salford, United Kingdom, and external collaborator of cE3c, University of Lisbon, Portugal. He holds an MSc degree from the University of Würzburg and a PhD from the University of Ulm (Germany).

His research centres on investigating the effects of habitat fragmentation and land-use change on tropical bats, and he has a considerable track record of conducting ecological and conservation-related studies on bats in various tropical countries, including Brazil.

Blanca Martí de Ahumada

Blanca graduated in History of Art from the University of Barcelona in 2004, specializing in illustration in 2003–2006 in the Francesca Bonnemaison school. Her keen interest in nature and animal biology led her to scientific illustration and enrollment on several courses at the Universities of Barcelona and Valencia, and the Galanthus Association. Carles Puche and Rosa Vidal are her two main mentors.

After publishing a number of books and many popular science articles, she is currently working as an art teacher on master courses on Ethology and Primatology at the Universities of Girona and Cordoba, and on courses at the Mona Foundation, University of Granada, and the Eventur Darwin and Sigantus Associations.

Eva Sánchez Gómez

Eva graduated in Fine Arts from the University of Barcelona in 2009, specializing in illustration for four years in the Francesca Bonnemaison school: Here, she illustrated her first album, *Onades i Flors*, followed by many other children's and poetry books, such as, most recently, *L'Attente*, published by Âne Bâté Éditions. Her work has been exhibited in several collective exhibitions, and was selected for the IV Ibero-American Catalogue in 2013 and Hipermerc'Art, an exhibition and market of contemporary art, for the last four editions (2011–2014),

She is greatly interested in science and nature, which influence much of her work.

References

MORPHOLOGICAL KEY

MAIN REFERENCE: Lim, B. & Engstrom, M. (2001) Species diversity of bats (Mammalia: Chiroptera) in Iwokrama Forest, Guyana, and the Guianan subregion: implications for conservation. *Biodiversity & Conservation* 10:613-657.

Aguirre, L.F. & Anderson, S. (2009) *Clave de campo para la identificación de los murciélagos de Bolivia.* Centro de Estudios en Biología Teórica y Aplicada, Cochabamba, Bolivia.

Bernard, E. et al. (2011) Compilação atualizada das espécies de morcegos (Chiroptera) para a Amazônia Brasileira. *Biota Neotropica* 11(1):35-46.

Charles-Dominique, P. (2001) *Les chauve-souris de Guyane.* Muséum National d'Histoire naturelle. Laboratoire d'Écologie générale. París. France.

Gardner, A.L. et al. (2007) *Mammals of South America.* Volume I: *Marsupials, Xenarthrans, Shrews, and Bats.* University of Chicago Press, Chicago and London.

Gregorin, R. & Taddei, V.A. (2002) Chave artificial para a identificação de molossídeos brasileiros (Mammalia, Chiroptera). *Mastozoologia Neotropical* 9(1):13-32.

López-González, C. (2005) *Murciélagos del Paraguay.* Biosfera, Madrid.

Medellin, R.A. et al. (1997) *Identificación de los murciélagos de México. Clave de campo.* Instituto de Ecología, Universidad Nacional Autónoma de México, Mexico City.

Miranda, J.M.D. et al. (2011) *Chave ilustrada para determinação dos morcegos da Região Sul do Brasil.* Laboratório de Biodiversidade, Conservação e Ecologia de Animais Silvestres (UFPR), Curitiba.

Reis, N. (2007) *Morcegos do Brasil.* UNESP, Universidade Estadual de Londrina, Londrina.

Reis N. et al. (2013) *Morcegos do Brasil. Guia de Campo.* Technical Books Editora, Rio de Janeiro.

SPECIFIC FAMILIES, GENERA AND SPECIES

Aires, C. (2008) Caracterização das espécies brasileiras de *Myotis* Kaup, 1829 (Chiroptera: Vespertilionidae) e ensaio sobre filogeografia de *Myotis nigricans* (Schinz, 1821) e *Myotis riparius* Handley, 1960. PhD thesis, Universidade de São Paulo, São Paulo.

Genoways, H.H. et al. (1986) Bats of the genus *Micronycteris* (Mammalia: Chiroptera) in Suriname. *Annals of Carnegie Museum* 55(13):303–324.

Gregorin, R. et al. (2006) New species of disk-winged bat *Thyroptera* and range extension for *T. discifera. Journal of Mammalogy* 87(2):238–246.

Nogueira, M.R. et al. (2014) Checklist of Brazilian bats, with comments on original records. *Check List* 10(4):808–821.

Porter, C.A. et al. (2007) Molecular phylogenetics of the phyllostomid bat genus *Micronycteris* with descriptions of two new subgenera. *Journal of Mammalogy* 88(5):1205–1215.

Siles, L. et al. (2013) A new species of *Micronycteris* (Chiroptera: Phyllostomidae) from Bolivia. *Journal of Mammalogy* 94(4):881–896.

Simmons, J.B. et al. (2002) A new Amazonian species of *Micronycteris* (Chiroptera : Phyllostomidae) with notes on the roosting behavior of sympatric congeners. *American Museum Novitates* 3358:1–14.

Velazco, P.M. & Patterson, B. (2014) Two new species of yellow-shouldered bats, genus *Sturnira* Gray, 1842 (Chiroptera, Phyllostomidae) from Costa Rica, Panama and western Ecuador. *ZooKeys* 402:43–66.

Velazco, P.M. et al. (2014) Extraordinary local diversity of disk-winged bats (Thyropteridae: *Thyroptera*) in northeastern Peru, with the description of a new species and comments on roosting behavior. *American Museum Novitates* 3795:1–28.

William, S.L. et al. (1995) Review of the *Tonatia bidens* complex (Mammalia: Chiroptera), with descriptions of two new subspecies. *Journal of Mammalogy* 76(2):612–626.

BIOACOUSTIC KEY

Audet, D. et al. (1993) Morphology, karyology, and echolocation calls of Rhogeessa (Chiroptera: Vespertilionidae) from the Yucatán Peninsula. *Journal of Mammalogy* 74(2):498–502.

Barataud, M. et al. (2013) Identification et écologie acoustique des chiroptères de Guyane française. *Le Rhinolophe* 19:43.

Bayefsky-Anand, S. (2006). Echolocation calls of the greater sac-winged bat (*Saccopteryx bilineata*) in different amounts of clutter. *Bat Research News* 47(1):4.

Clare, E.L. et al. (2013) Diversification and reproductive isolation: cryptic species in the only New World high-duty cycle bat, *Pteronotus parnellii. BMC Evolutionary Biology* 13(26):1–18.

Clement, M.J. et al. (2014) The effect of call libraries and acoustic filters on the identification of bat echolocation. *Ecology and Evolution* 4(17):3482–3493.

Cormier, A.C.A. (2014) Species diversity and activity of insectivorous bats in three habitats in La Virgen de Sarapiquí, Costa Rica. *Revista De Biologia Tropical* 62(3):939–946.

Fenton, M.B. & Bell, G.P. (1981) Recognition of species of insectivorous bats by their echolocation calls. *Journal of Mammalogy* 62(2):233–243.

Fenton, M.B. et al. (1999) Constant-frequency and frequency-modulated components in the echolocation calls of three species of small bats (Emballonuridae, Thyropteridae, and Vespertilionidae). *Canadian Journal of Zoology* 77(12):1891–1900.

Gillam, E.H. & Chaverri, G. (2012) Strong individual signatures and weaker group signatures in contact calls of Spix's disc-winged bat, *Thyroptera tricolor*. *Animal Behaviour* 83(1):269–276.

Griffin, D.R. & Novick, A. (1955) Acoustic orientation of neotropical bats. *Journal of Experimental Zoology* 130(2):251–299.

Heer, K. et al. (2015) Effects of land use on bat diversity in a complex plantation-forest landscape in northeastern Brazil. *Journal of Mammalogy* 96(4):720–731.

Ibáñez, C. et al. (1999) Echolocation calls of *Pteronotus davyi* (Chiroptera: Mormoopidae) from Panama. *Journal of Mammalogy* 80(3):924–928.

Jung, K. et al. (2007) Echolocation calls in Central American emballonurid bats: signal design and call frequency alternation. *Journal of Zoology* 272(2):125–137.

Jung, K. & Kalko, E.K. (2010) Where forest meets urbanization: foraging plasticity of aerial insectivorous bats in an anthropogenically altered environment. *Journal of Mammalogy* 91(1):144–153.

Jung, K. & Kalko, E.K. (2011) Adaptability and vulnerability of high flying neotropical aerial insectivorous bats to urbanization. *Diversity and Distributions* 17(2):262–274.

Jung, K. et al. (2014). Driving factors for the evolution of species-specific echolocation call design in New World free-tailed bats (Molossidae). *PLoS ONE* 9(1):e85279.

López-Baucells, A. et al. (2014) Echolocation of the big red bat *Lasiurus egregius* (Chiroptera: Vespertilionidae) and first record from the Central Brazilian Amazon. *Studies on Neotropical Fauna and Environment* 49(1):18–25.

López-Baucells, A. et al. (2017) Molecular, morphological and acoustic identification of *Eumops maurus* and *Eumops hansae* (Chiroptera: Molossidae) with new reports from Central Amazonia. *Tropical Zoology*.

López-Baucells, A. et al. (2017) Geographical variation in the high-duty cycle echolocation of the cryptic common mustached bat *Pteronotus* cf. *rubiginosus* (Mormoopidae). *Bioacoustics Journal*.

MacSwiney, G.M.C. et al. (2009) Insectivorous bat activity at Cenotes in the Yucatan Peninsula, Mexico. *Acta Chiropterologica* 11(1):139–147.

Miller, B. W. (2003) Community ecology of the non-phyllostomid bats of northwestern Belize, with a landscape level assessment of the bats of Belize. PhD thesis, Institute of Conservation and Ecology, University of Kent, Canterbury.

Moratelli, R. & Oliveira, J.A.D. (2011) Morphometric and morphological variation in South American populations of *Myotis albescens* (Chiroptera: Vespertilionidae). *Zoologia* 28(6):789–802.

Novick, A. & Vaisnys, J.R. (1964) Echolocation of flying insects by the bat, *Chilonycteris parnellii*. *Biological Bulletin* 127(3):478–488.

O'Farrell, M.J. & Miller, B.W. (1997) A new examination of echolocation calls of some neotropical bats (Emballonuridae and Mormoopidae). *Journal of Mammalogy* 78(3):954–963.

O'Farrell, M.J. et al. (1999) Qualitative identification of free-flying bats using the Anabat detector. *Journal of Mammalogy* 80(1):11–23.

Rodhouse, T.J. et al. (2011) A practical sampling design for acoustic surveys of bats. *Journal of Wildlife Management* 75(5):1094–1102.

Rodriguez, A. et al. (2014) Temporal and spatial variability of animal sound within a neotropical forest. *Ecological Informatics* 21:133–143.

Rodríguez, A. & Rodríguez, A. (2011) Acoustic identification of *Nycticeius cubanus* (Gundlach, 1867) and *Eptesicus fuscus dutertreus* (Gervais, 1837) (Chiroptera: Vespertilionidae) in western Cuba. *Revista Cubana de Ciencias Biológicas* 19:1–2.

Rodríguez, A. & Mora, E.C. (2006) The echolocation repertoire of *Eptesicus fuscus* (Chiroptera: Vespertilionidae) in Cuba. *Caribbean Journal of Science* 42(1):121.

Rodríguez-San Pedro, A. & Simonetti, J.A. (2013). Acoustic identification of four species of bats (Order Chiroptera) in central Chile. *Bioacoustics* 22(2):165–172.

Rodríguez-San Pedro, A. & Simonetti, J.A. (2014) Does understory clutter reduce bat activity in forestry pine plantations? *European Journal of Wildlife Research* 61(1):177–179.

Rydell, J. et al. (2002) Acoustic identification of insectivorous bats (order Chiroptera) of Yucatan, Mexico. *Journal of Zoology* 257(1):27–36.

Schnitzler, H. et al. (1991) Comparative studies of echolocation and hunting behaviour in the four species of mormoopid bats of Jamaica. *Bat Research News* 32:22–23.

Siemers, B. et al. (2001) Echolocation behavior and signal plasticity in the neotropical bat *Myotis nigricans* (Schinz, 1821) (Vespertilionidae): a convergent case with European species of *Pipistrellus*? *Behavioral Ecology and Sociobiology* 50(4):317–328.

Thoisy, B.D. et al. (2014) Cryptic diversity in common mustached bats *Pteronotus* cf. *parnellii* (Mormoopidae) in French Guiana and Brazilian Amapa. *Acta Chiropterologica* 16(1):1–13.

BDFFP Camp in the central Amazon

Collaborating institutions

UNIVERSIDADE
FEDERAL
DE PERNAMBUCO

P D B F F
I N P A · S I
Projeto Dinâmica Biológica
de Fragmentos Florestais

INPA
INSTITUTO NACIONAL DE
PESQUISAS DA AMAZÔNIA

editoraINPA

Ministério da
**Ciência Tecnologia,
Inovações e Comunicações**

UNIVERSITY OF HELSINKI

UNIVERSIDADE da MADEIRA

Index